Introductory
Applied Statistics
In Science

Sung C. Choi

Washington University
St. Louis

Prentice-Hall, Inc., Englewood Cliffs, New Jersey   07632

Library of Congress Cataloging in Publication Data

CHOI, SUNG C
   Introductory applied statistics in science.

   Bibliography: p.
   Includes index.
   1. Mathematical statistics. 2. Biometry.
3. Science—Statistical methods. I. Title.
QA276.C474   519.5'02'45   77–22869
ISBN 0–13–501619–3

© 1978 by Prentice-Hall, Inc., Englewood Cliffs, N.J. 07632

Printed in the United States of America

10 9 8 7 6 5 4 3 2 1

Prentice-Hall International, Inc., *London*
Prentice-Hall of Australia Pty. Limited, *Sydney*
Prentice-Hall of Canada, Ltd., *Toronto*
Prentice-Hall of India Private Limited, *New Delhi*
Prentice-Hall of Japan, Inc., *Tokyo*
Prentice-Hall of Southeast Asia Pte. Ltd., *Singapore*
Whitehall Books Limited, *Wellington, New Zealand*

Contents

iii

Preface

The outline of this book evolved from lecture notes and materials, which were used in one-semester courses in applied statistics in the School of Engineering and Applied Science, and in several biostatistics courses in the School of Medicine at Washington University, St. Louis, Missouri.

It is designed to be a text in elementary applied statistics for students in science, mainly engineering and biomedical fields, and a compact reference text for the applied statistician and researcher. As a textbook it can be covered in a single semester if Sec. 9.6–9.10, Chap. 10, together with about eight additional sections, at the discretion of instructors, are omitted. Chap. 9 and 10 deal with relatively more complex topics, and are presented primarily for additional reading and reference. The prerequisite of this book is a standard algebra course and some of the basic ideas of calculus. Those readers without an introduction to calculus can omit the parts where calculus is used without loss of continuity.

In writing this book, an attempt was made to motivate the readers with the applicability and usefulness of statistical methods in the real world. Throughout the book, the methods are illustrated with real or realistic examples. Some slant toward medical examples perhaps reflects the author's interest and the availability of data. In addition, special sets of problems, many of them based on real data, follow each of the first nine chapters. It is hoped that those materials most frequently required or useful in applications are included here, though to some extent the importance of the materials depends on the field of applications. Although it is an applied statistics text, some effort is made to give those interested readers the rationality of the methods wherever possible.

It is suggested that the starred sections and starred problems be omitted for a shorter term course or for a strictly application-oriented course. In addition, the following alternative sequences come to mind for even shorter courses or for readers with different backgrounds and different interests.

    (a) Chapter 1 (1.1 through 1.4), Chap. 3, Chap. 5 (5.1, 5.5, 5.6, 5.8), Chap. 6 through Chap. 9.

    (b) Chapter 4, Sec. 3.6, Chap. 5 (5.1, 5.5, 5.6, 5.8), Chap. 6 through Chap. 9.

    (c) Chapter 5 (5.1, 5.5, 5.6, 5.8), Chap. 6 through Chap. 9.

The author is indebted to various publishers for permission to reproduce and adopt tables and figures as acknowledged in each table and figure. He is grateful to the Literary Executor of the late Sir Ronald A. Fisher, F.R.S., to Dr. Frank Yates, F.R.S., and to Longman Group Ltd., London, for permission to reprint Table III from their book, *Statistical Tables for Biological, Agricultural and Medical Research*, (6th edition, 1974).

He also wishes to express his appreciation to Sara Sanders for preparing the manuscript, and to many unnamed individuals, but particularly to Barbara Hixon for making helpful comments and suggestions. Finally, the author would like to thank the editorial staff of Prentice-Hall, Inc. for their friendly cooperation in the production of this book.

St. Louis, Missouri   SUNG C. CHOI

Introduction

Statistics is the scientific method of collecting information in a form of numerical data and drawing conclusions by analyzing the information. Consider, for example, the following problems:

1. deciding whether or not a certain game is fair;
2. estimating the number of fish in a lake;
2. determining the unemployment rate;
4. deciding whether or not a drug is effective;
5. comparing mileage obtained using several different brands of gasoline;
6. testing the possible relation between the length of the "life-line" on the hand and life expectancy;
7. deciding whether or not cigarette smoking causes cancer; or
8. estimating the yield of wheat for different amounts of a standard fertilizer applied.

In each of the above problems, the only practical scientific approach is to perform some sort of experiment or survey and base the solution on the information obtained. But what kind of information and how much? And after we have the information, what do we do with it to solve the problem? Statistics deals with answering these kinds of questions by specific techniques.

Statistics usually consists of four broad processes, although there are not always clear boundaries between them: collection, organization, analysis of numerical data, and the decision process.

Collecting data is the process of obtaining measurements or counts after some sort of experiment or survey is conducted. Valid conclusions can result only from properly collected data.

Organization of data is the process of preparing and presenting the collected data in a form suitable for description as well as for further analysis.

Analysis of data is the process of performing certain calculations and evaluations in order to extract relevant and pertinent information buried in the data.

The decision process is the task of interpreting and reaching valid conclusions based on the analysis of the data and the mathematical theory of probability. The analysis of data and the decision process form the main portion of this book.

1

Random Variables and Probability

1.1 Introduction

In many scientific studies, we deal with experiments that are repetitive in nature or that can be conceived as being repetitive. For example, if we toss a coin ten times, we may want to know the chance that no head will appear, one head will appear, two heads will appear, and so on. If a sample of 100 electronic tubes is selected from a shipment and each tube is tested, it may be desirable to estimate the proportion of defective tubes in the lot. In a medical investigation, if the survival time of a group of sick mice receiving a placebo and a second group receiving a certain medication is recorded, we may want to examine the effectiveness of the treatment in terms of survival time. These are illustrations of experiments that can be carried out actually or conceptually. In the study of probability, we are concerned with the derivation of the laws and rules of chance related to the outcomes of experiments.

1.2 Sample Space, Random Variable, and Probability

Consider an experiment of tossing a coin. In the experiment, there are only two possible outcomes, a head or a tail. It is convenient to represent head and tail by certain letters, for example, H for head and T for tail, although such letters can be arbitrary. For an experiment of rolling a die, there are six possible outcomes, most conveniently represented as $1, 2, \ldots, 6$.

A point representing a possible outcome of a given experiment is called a *sample point*, and the set of all sample points is called the *sample space.* A numerically defined variable on a sample space is called a *random variable*. To be precise, a random variable is a numerical function defined on each element of the sample space. In practice, however, a random variable shall be conceived as a numerical value assigned to each element.

3

To be precise, let $S = \{e_1, e_2, \ldots, e_n\}$ denote the sample space, with each e_i representing a sample point. A random variable X is defined by assigning a real number x_i to each element e_i of S. Thus, we may write

$$X(e_i) = x_i.$$

In practice, the random variable defined by an investigator depends on the nature and purpose of the study and the criterion used, and it is usually denoted simply by X instead of $X(e_i)$. Also, it is often called the *variable* for the sake of simplicity.

Example 1.1 Consider the experiment of rolling a die. The sample points can be conveniently represented by $1, 2, \ldots, 6$, and the sample space is given by a set $\{1, 2, \ldots, 6\}$. A random variable X can be defined as

$$X(1) = 1, \ X(2) = 2, \ldots, X(6) = 6.$$

The relationship between the outcome and the random variable defined here is clearly illustrated in Fig. 1.1.

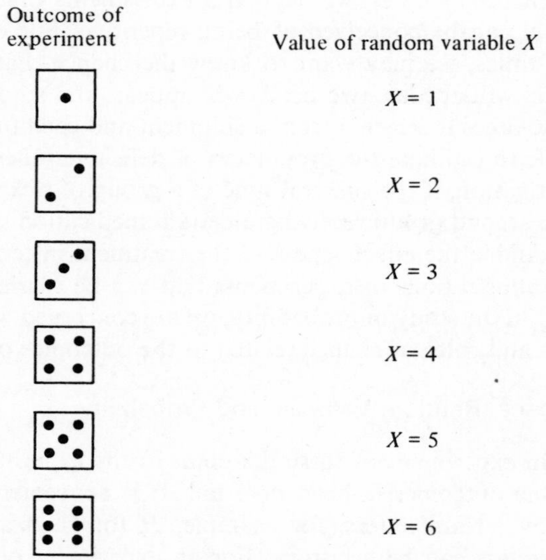

Outcome of
experiment   Value of random variable X

$X = 1$

$X = 2$

$X = 3$

$X = 4$

$X = 5$

$X = 6$

Fig. 1.1   Relationship between outcome and random variable

Alternatively, X can be defined as follows if one is concerned only as to the outcome being an odd or an even number.

$$X(1) = X(3) = X(5) = 0, \ \text{and} \ X(2) = X(4) = X(6) = 1,$$

or in many other ways. ▲

Example 1.2 As the second example, a random variable might be defined as the number of heads appearing when two coins are tossed. Then, $S = \{HH, HT, TH, TT\}$, and

$$X(H, H) = 2, \ X(H, T) = X(T, H) = 1, \text{ and } X(T, T) = 0. \ \blacktriangle$$

Example 1.3 Consider the survival time of mice with a certain disease. The survival time can be any positive real number as is the sample point. The sample space is continuous and is given by a set $\{t \mid 0 \leq t < T\}$, where T is a large real number. A random variable X can be defined as the survival time of the animal itself; that is,

$$X(t) = t.$$

Alternatively, it can be defined as

$$X(t) = 0 \text{ if } 0 \leq t \leq 5 \text{ days,}$$
$$X(t) = 1 \text{ if } 5 < t \leq 18 \text{ days,}$$
$$X(t) = 2 \text{ if } t > 18 \text{ days,}$$

and again in many other ways. ▲

Note that each outcome always determines one and only one value of the random variable X, but a given value of X may correspond to more than one outcome, although often assigned to only one outcome. Indeed, in many situations, the outcome of the experiment is already in the numerical form that we want to record and use as a random variable.

Returning to the sample space of an experiment, any subset of the sample space is called an *event*. The events shall be noted by capital letters, for example, A, B, etc. Suppose that the experiment related to the event A is performed. Every sample point belongs to either A or not to the event A. Only if the sample point belongs to A, can the event A then be said to have occurred. For example, for the experiment of rolling a die, let the event be defined as $A = \{2, 4, 6\}$. Then, we say that the event A has occurred when each roll of the die results in one of the three numbers, namely, 2, 4, and 6.

Given any event A, it is natural to consider the event that A does not occur, denoted by $\bar{A}$. Such an event is called the *complement* of A and consists of all elements in the sample space which are not in A. For example, if $A = \{2, 4, 6\}$ in the experiment of rolling a die, then $\bar{A} = \{1, 3, 5\}$.

As has been stated, the random variable, or simply the variable, is a number assigned to the outcome of an experiment and, as might be expected, there is a relation between the variable and an event. In brief, the variable can define an event. Thus, sample points determined by $\{a \leq X \leq b\}$, where X is a variable and a and b are real numbers, always constitute an event. For example, $\{X = 2\}$ in Example 1.2 determines the event that both coins show heads, and $\{0 \leq X \leq 1\}$ the event characterized by "at most one head." In

Example 1.3, if the variable X is defined by $X = t$, then $\{X > 3\}$ defines the event that the survival time of the animal is greater than three days, if day was the time scale. In statistical analyses, we shall deal with the observed values of X or the event defined by X. The main reason for, and advantage in, defining the random variable should be clear to the reader: it is much more convenient to work with a set of given numbers precisely defined on the outcome of an experiment or observation than to work with the outcome itself.

Example 1.4 Consider the weight of couples. Let X and Y denote the weights of a husband and wife, respectively. Each sample point is given by (X, Y), $X > 0$, $Y > 0$, although in reality X and Y are bound by certain values, and the sample space can be represented by the first quadrant of the X, Y-plane. The event A, "wife is heavier than husband," for example, is given by the shaded region of Fig. 1.2.

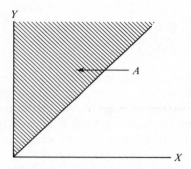

Fig. 1.2 Event A "Wife is heavier than husband": X is weight of husband and Y the weight of wife

Many random variables can be defined; for instance, the random variable W can be the sum of a couple's weights, that is, $W = X + Y$. Next, the random variable D can be defined as the absolute difference between the weights, $|X - Y|$. It is also true that both X and Y are random variables. As a final example, let $Z = 1$ if $X < Y$ and $Z = 0$ if $X \geq Y$. Then, Z is clearly a random variable, and $\{Z = 1\}$ is characterized by the event A of Fig. 1.2. ▲

A basic and intuitive meaning of probability is most easily given when the sample space consists of a finite number of sample points, and when each sample point is equally likely to occur in a repetitive experiment. Suppose we are interestsd in a certain event A as to the likelihood that the event will occur in a single trial. The classical definition states: the *probability* of an event A is the ratio of the number of sample points in A to the total number of sample points. Thus, if $P(A)$ denotes the probability of the event A, and if

n_A and n denote the number of sample points in A and the total number of sample points respectively, then

$$P(A) = \frac{n_A}{n}. \tag{1.1}$$

For example, consider again the experiment of rolling a fair die with the event A defined as $A = \{2, 4, 6\}$. Since each of six numbers is equally likely to appear in the experiment, $P(A) = \frac{3}{6} = \frac{1}{2}$. Note that if a random variable is defined as $X(1) = 1$, $X(2) = 2, \ldots, X(6) = 6$, then $P(A) = P(X = 2, 4, \text{ or } 6)$. On the other hand, if it is defined as $X(1) = X(3) = X(5) = 0$ and $X(2) = X(4) = X(6) = 1$, then $P(A) = P(X = 1)$.

If the total number of sample points is infinite, the definition given by (1.1) is, of course, not appropriate. More generally, we may define the probability of the event A as the relative chance or likelihood that A occurs in a given experiment. This means roughly that the probability is the fraction of times the event A occurs if the experiment is repeated a large number of times under essentially identical conditions. This definition is somewhat ambiguous, but the more exact and broad meaning of the probability will become clear in the next chapter.

1.3 Addition Theorem

Suppose we have two events A and B. The event consisting of all sample points contained in A or B, or both, is called the *union* of A and B: it is written as

$$A \cup B.$$

The event consisting of all points contained in both A and B is called the *intersection* of A and B: it is written

$$A \cap B.$$

In Fig. 1.3 the event $A \cup B$ is represented by the unshaded plus the shaded regions of A and B, while the event $A \cap B$ is given by the shaded region only. Such a figure is known as a Venn diagram.

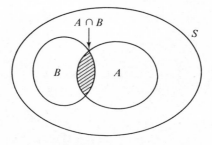

Fig. 1.3 Two events A and B within sample space S

Let $P(A)$ and $P(B)$ be the probabilities of two events, A and B, respectively. The probability that either A or B or both occur is denoted by $P(A \cup B)$. In other words, $P(A \cup B)$ is the probability of occurrence of at least one of the two events. Similarly, $P(A \cap B)$ denotes the probability that both A and B occur together. The addition theorem gives the rule of computing $P(A \cup B)$ from $P(A)$, $P(B)$, and $P(A \cap B)$. The _addition theorem_ is given by

Theorem 1A

$$P(A \cup B) = P(A) + P(B) - P(A \cap B). \qquad (1.2)$$

The addition of the probabilities $P(A)$ and $P(B)$ results in including the points of $A \cap B$ twice; hence, $P(A \cap B)$ must be subtracted once from the sum $P(A) + P(B)$. Equation (1.2) is also called the either-or theorem of probability.

Theorem 1A can be extended to three or more events as follows. If A, B, and C are any three events, then as we can show by the Venn diagram, the probability of the event $A \cup B \cup C$ is given by

$$P(A \cup B \cup C) = P(A) + P(B) + P(C) - P(A \cap B) - P(A \cap C)$$
$$- P(B \cap C) + P(A \cap B \cap C). \qquad (1.3)$$

If A and B are disjointed events so that $A \cap B = \varnothing$, they cannot occur simultaneously; they are defined to be _mutually exclusive_ events. Two mutually exclusive events are shown in the Venn diagram of Fig. 1.4. Note that the two regions A and B have no points in common. For example, let A be the event that one is a male and B that one is a female. Then, clearly, A and B are mutually exclusive because no one can be both male and female. On the other hand, for example, having cancer and having heart disease are not mutually exclusive since one can have both of these two diseases. The definition extends to three or more events.

If the two events A and B are mutually exclusive, then occurrence of the event $A \cap B$ is impossible, and thus

$$P(A \cap B) = 0. \qquad (1.4)$$

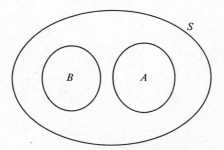

Fig. 1.4 Two mutually exclusive events A and B within sample space S

In this case, the addition theorem reduces to a simpler form, sometimes called the theorem of total probability;

$$P(A \cup B) = P(A) + P(B). \tag{1.5}$$

Example 1.5 In a public health survey, heart disease was found in 9% of a certain population. The same survey showed that hypertension was present in 12% of the individuals. Furthermore, 7% of the examined individuals had both heart disease and hypertension. Suppose it is desired to find the proportion of individuals in the population with neither condition. Let D and H denote the events characterizing heart disease and hypertension, respectively. Then, it is given $P(H) = 0.12$, $P(D) = 0.09$, and $P(H \cap D) = 0.07$. From Eq. (1.2) the proportion of either condition is given by

$$P(H \cup D) = 0.12 + 0.09 - 0.07 = 0.14.$$

Since neither condition is the complement of either condition, the desired proportion is given by

$$1 - P(H \cup D) = 0.86. \quad \blacktriangle$$

1.4 Conditional Probability, Independent Events, and Multiplication Theorem

For two given events A and B the probability of the event B occurring if one knows that A occurs is defined as *conditional probability* of B given A and denoted as $P(B|A)$. In everyday life we are concerned with the concept of the conditional probability perhaps without realizing it. For example, we may be concerned with "chance of snow when it is freezing," "chance of winning a poker hand given that one has a pair of kings," "chance of curing a cancer patient if a treatment is given," and so on.

If $P(A \cap B)$ and $P(A)$ are known, the conditional probability of B given A can be calculated by the formula

$$P(B|A) = \frac{P(A \cap B)}{P(A)}. \tag{1.6}$$

Equation (1.6) makes sense if we look at Fig. 1.3, since $P(B|A)$ can be interpreted as the probability that B will occur knowing that the event A occurs. Conditional probabilities are undefined if $P(A) = 0$.

If the occurrence of the event B is not influenced or dependent on a second event A, for which $P(A) \neq 0$ and $P(B) \neq 0$, so that

$$P(B|A) = P(B),$$

then A and B are said to be *independent*.

Formula (1.6) is often used in the form known as the following *multiplication theorem*.

Theorem 1B

$$P(A \cap B) = P(A) P(B \mid A). \tag{1.7}$$

This theorem may be generalized to three events as follows:

$$P(A \cap B \cap C) = P(A) P(B \mid A) P(C \mid A \cap B).$$

If A and B are independent, then the multiplication theorem given by (1.7) reduces to

$$P(A \cap B) = P(A) P(B), \tag{1.8}$$

and the addition theorem given by (1.2) becomes

$$P(A \cup B) = P(A) + P(B) - P(A) P(B). \tag{1.9}$$

Example 1.6 Consider Example 1.5 dealing with heart disease and hypertension of a population. (a) What is the probability that an individual has heart disease if he has hypertension? (b) Are the two conditions independent?
It was given $P(H) = 0.12$, $P(D) = 0.09$, and $P(H \cap D) = 0.07$. To answer the first question, we use (1.6) to obtain

$$P(D \mid H) = \frac{0.07}{0.12} = 0.58.$$

To answer the second question, we see that $P(D \mid H) \neq P(D)$. Thus, the two conditions are not independent. The nonindependency of the conditions is also checked by $P(H) P(D) \neq P(H \cap D)$ using (1.8). ▲

The dependency of the two events H and D as shown in Example 1.6 was anticipated since heart disease and hypertension are known to be related. Similarly, sometimes independence of events is obvious. For example, suppose that a fair coin and a die are tossed together, and let A denote the event of getting a head on the coin and B that of getting six on the die. The two events are clearly independent because the occurrence of A and B do not influence each other. It follows that $P(A \mid B) = P(A) = \frac{1}{2}$ and $P(A \cap B) = P(A) P(B) = \frac{1}{12}$.

The independence or dependence of events, however, is not always apparent until verified. As an example, consider families with two children. Letting b represent boy and g, girl, the sample space is given by $\{bb, bg, gb, gg\}$ where the first letter stands for the older child. Let A be the event "the family has children of both sexes," and B the event "one or no boys in the family." Using Eq. (1.1), we have $P(A) = \frac{2}{4}$, $P(B) = \frac{3}{4}$, and since $A \cap B$ represents the event "exactly one boy," $P(A \cap B) = \frac{2}{4}$. Thus, A and B are not independent. Next, consider families with three children. The sample space, again differentiating the order of children, is given by $\{bbb, bbg, bgb, gbb, bgg, gbg, ggb, ggg\}$. Letting A and B represent the same events as above, we have $P(A) = \frac{6}{8}$, $P(B) = \frac{4}{8}$, and $P(A \cap B) = \frac{3}{8}$. Thus, in families with three children the two events are shown to be independent. The reader may wish to show that the two events A and B defined for families with four children are not independent.

Incidentally, mutually exclusive events should not be confused with independent events. Two mutually exclusive events A and B are necessarily dependent. For, if A occurs, then B cannot occur, if they are mutually exclusive, so that in general if A and B are mutually exclusive, then

$$P(A \mid B) = 0 \neq P(A),$$

and, similarly,

$$P(B \mid A) = 0 \neq P(B).$$

Thus, mutually exclusive events are not independent. The converse, however, is false. For example, the two events H and D in Example 1.6 are not independent, but they are not mutually exclusive; that is, $H \cap D \neq \varnothing$.

In view of the relation between the event and the random variable, we would expect the ideas of independence to be extended to the variables. Two variables X and Y are said to be *independent* if

$$P(X < a, \ Y < b) = P(X < a) \, P(Y < b)$$

for any constants a and b. As a simple example, let the variable X be defined as 1 or 0 depending on whether a toss of a coin shows a head or a tail, Y be the number showing when a die is rolled, and at the same time, the variable Z be defined as 1 if this die shows an even number, and 0 otherwise. Then, it it is easy to see that X is independent of Y or Z, whereas Y and Z are not independent.

Before leaving this section, we shall note the following rule on conditional probability which is often found to be useful. The rule is similar to a simpler relation, $P(\bar{A}) = 1 - P(A)$ where A is an event.

Theorem 1C

For two given events A and B,

$$P(A \mid B) = 1 - P(\bar{A} \mid B).$$

Theorem 1C can be easily verified by showing that $P(A \mid B) + P(\bar{A} \mid B) = 1$, using the definition given by Eq. (1.6) and a Venn diagram like Fig. 1.3. The above result should not be confused with the inequality $P(A \mid B) \neq 1 - P(A \mid \bar{B})$.

Example 1.7 In Example 1.5, suppose that 1% of individuals had heart disease but had normal blood pressure. What is the probability that a randomly selected individual from a group with normal blood pressures has no heart disease?

As in Example 1.5, let D and H represent the events of an individual having heart disease and having hypertension, respectively. Then, it is given that $P(D \cap \bar{H}) = 0.01$. Since $P(H) = 0.12$, we calculate that $P(\bar{H}) = 1 - P(H) = 0.88$. It follows that

$$P(D|\bar{H}) = \frac{P(D \cap \bar{H})}{P(\bar{H})} = \frac{0.01}{0.88} = 0.011.$$

The answer to the question is calculated next using Theorem 1C as

$$P(\bar{D}|\bar{H}) = 1 - P(D|\bar{H}) = 1 - 0.011 = 0.989. \quad \blacktriangle$$

1.5 Theorem on Total Probability

For a sample space S if the union of two or more given events completely covers S, then those events are said to be *exhaustive* of S. Let B_1, $B_2, \ldots, B_k$ represent mutually exclusive and exhaustive events of the sample space S. That is,

(a) $B_i \cap B_j = \varnothing$ for all $i \neq j$,

(b) $B_1 \cup B_2 \cup \ldots \cup B_k = S$.

Let A be a certain event. For example, the diagram in Fig. 1.5 depicts such a situation for $k = 5$. We can express the event A as follows:

$$A = (A \cap B_1) \cup (A \cap B_2) \cup \ldots \cup (A \cap B_k).$$

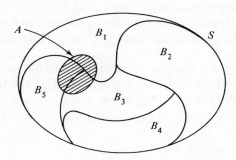

Fig. 1.5   Five mutually exclusive and exhaustive events

Of course, some of the sets $A \cap B_j$ may be empty, but the above expression still holds true. The point is that all the events $(A \cap B_1), \ldots, (A \cap B_k)$ are mutually exclusive. Hence, applying the rule for mutually exclusive events which is the generalization of (1.5), we have

$$P(A) = P(A \cap B_1) + P(A \cap B_2) + \cdots + P(A \cap B_k).$$

Moreover, from (1.7) we know that $P(A \cap B_j) = P(A|B_j)P(B_j)$, and hence we obtain a useful relationship known as the *theorem on total probability*:

$$P(A) = P(A|B_1)P(B_1) + P(A|B_2)P(B_2) \qquad (1.10)$$
$$+ \cdots + P(A|B_k)P(B_k).$$

Often $P(A)$ cannot be computed directly, but with the information on B_j its computation becomes an easy problem as illustrated in the following example.

Example 1.8 In a factory, machines A, B, C manufacture 25%, 35%, and 40% of the total output of a certain item, respectively. Of their outputs, 1%, 2%, and 2.5%, respectively, are known to be defective. Estimate the number of defective items in a lot of 10,000 items.

Let D denote the event that an item is defective. From (1.10) we have

$$P(D) = P(D\,|\,A)\,P(A) + P(D\,|\,B)\,P(B) + P(D\,|\,C)\,P(C)$$
$$= (0.01)(0.25) + (0.02)(0.35) + (0.025)(0.40) = 0.0195.$$

Thus, the probability of a randomly selected item being defective is 0.0195. Hence, on the average, $(10{,}000)(0.0195) = 195$ defective items are expected in the lot consisting of 10,000. ▲

1.6 Bayes' Theorem

Suppose that $P(B\,|\,A)$, the conditional probability of B given A is known, but we are really interested in knowing $P(A\,|\,B)$. The situation arises frequently in applications. For example, we can estimate the proportion of smokers among cancer patients; thus, we have the conditional probability of smoking, given that an individual has cancer. We can estimate the proportion of coffee drinkers among heart patients; thus, we have the conditional probability of coffee drinking, knowing that an individual has a heart disease. Clearly, we would be more interested in determining the conditional probability of one having cancer given that he is a smoker, and that of one having heart disease given that he is a coffee drinker.

At the outset we note that

$$P(A\,|\,B) \neq P(B\,|\,A)$$

unless $P(A) = P(B)$. Indeed, the two conditional probabilities can be very much different. The following simple example illustrates the point. Consider a population consisting of just ten individuals of which four have a certain disease and six do not have the disease. Suppose that a laboratory test was given to all the individuals. In Fig. 1.6, D and $\bar{D}$ represent the individuals with and without the disease, respectively, and $+$ and $-$ denote positive and negative test results, respectively. Thus, for example, $\bar{D}_+$ indicates an individual without the disease but with a positive test result.

From Fig. 1.6, the conditional probability $P(+\,|\,D)$ is the proportion of those with the disease having a positive test, and $P(D\,|\,+)$ the proportion of individuals with the disease among those with a positive test result. Thus,

$$P(+\,|\,D) = \tfrac{4}{4} = 1.0;$$

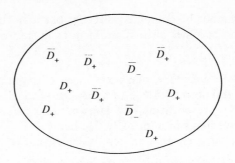

Fig. 1.6 A population of ten individuals

whereas,

$$P(D\,|+) = \tfrac{4}{8} = 0.5,$$

and $P(\bar{D}\,|+) = 0.5$. It implies that the test has a 50% chance of falsely indicating the disease even though all individuals with the disease give a correct indication. Similarly,

$$P(-\,|\,\bar{D}) = \tfrac{2}{6} = 0.333,$$
$$P(\bar{D}\,|-) = \tfrac{2}{2} = 1.0.$$

Incidentally, in medical literature, the conditional probability $P(+\,|\,D)$ is known as the *sensitivity* and $P(-\,|\,\bar{D})$ the *specificity* of the test.

The above illustration underscores the need and usefulness of having a formula which can be used to calculate the conditional probability $P(A\,|\,B)$ from $P(B\,|\,A)$ or vice versa. As mentioned already, it is a common problem in conditional probability when conditional probabilities are given "in the wrong direction." The formula which can be used in the situation is given in the following theorem.

Theorem 1D

For given $P(B\,|\,A)$, $P(A)$, $P(B\,|\,\bar{A})$, the conditional probability $P(A\,|\,B)$ is determined by the formula

$$P(A\,|\,B) = \frac{P(B\,|\,A)\,P(A)}{P(B\,|\,A)\,P(A) + P(B\,|\,\bar{A})\,P(\bar{A})}. \qquad (1.11)$$

Formula (1.11) is a special case of the result known as *Bayes' theorem.* The difficulty with the theorem is that in many circumstances the unconditional probability $P(A)$ is rarely known, not to mention the fact that $P(B\,|\,\bar{A})$ should also be known. Nevertheless, the theorem can be applied satisfactorily in some situations.

Example 1.9 Suppose that a diagnostic test for a certain disease is being developed. When tried on sick patients, the test gave positive results in 95% of the cases, and in healthy individuals it gave negative results for 90%. Let

D denote the event that a person tested has the disease, and $+$ and $-$ denote events that a test is positive or negative. We have $P(+\,|\,D) = 0.95$ and $P(-\,|\,\bar{D}) = 0.90$. Let us assume that the probability that a person taking the test actually has the disease is $P(D) = 0.01$.

In order to assess the reliability of the test, it is desirable to determine $P(D\,|\,+)$. Using (1.11) we can write

$$P(D\,|\,+) = \frac{P(+\,|\,D)\,P(D)}{P(+\,|\,D)\,P(D) + P(+\,|\,\bar{D})\,P(\bar{D})}.$$

Since $P(+\,|\,\bar{D}) = 1 - P(-\,|\,\bar{D})$ and $P(\bar{D}) = 1 - P(D)$, we have

$$P(D\,|\,+) = \frac{(0.95)(0.01)}{(0.95)(0.01) + (0.10)(0.99)} = 0.088.$$

Thus, even if the test gives a positive result, the chance is only 8.8% that the person has the disease. In other words, since $P(\bar{D}\,|\,+) = 1 - P(D\,|\,+)$ the chance of the false positive by the test is 91.2%. ▲

Formula (1.11) can be generalized to the following form which is known as Bayes' theorem.

Theorem 1E (*Bayes' Theorem*)

Let $A_1, A_2, \ldots, A_k$ be k mutually exclusive and exhaustive events, and let B be an event for which one knows the k conditional probabilities $P(B\,|\,A_i)$ and also the k unconditional probabilities $P(A_i)$. Then, we have

$$P(A_i\,|\,B) = \frac{P(B\,|\,A_i)\,P(A_i)}{\sum\limits_{j=1}^{k} P(B\,|\,A_j)\,P(A_j)}. \qquad (1.12)$$

Example 1.10 A certain item is manufactured by three companies, say 1, 2, and 3. It is known that the proportions of the items in a lot produced by the three companies are 50%, 25%, and 25%, respectively, for 1, 2, and 3. It is also known that 2% of the items produced by 1 and by 2 are defective, while 1% by 3 are defective. Suppose that one item is randomly chosen from the lot and is found to be defective. What is the probability that it was manufactured by 1?

Let us define the following events:

$B = \{$the item is defective$\}$, $A_i = \{$the item came from i; $i = 1, 2, 3\}$.

Using (1.12), the desired probability given by $P(A_1\,|\,B)$ is

$$P(A_1\,|\,B) = \frac{(0.02)(0.5)}{(0.02)(0.5) + (0.02)(0.25) + (0.01)(0.25)} = 0.57.$$

So the answer is 57%. ▲

1.7 Summary

A brief resume of both verbal and symbolic notations and the formulas used in some simple manipulations of the probability is summarized in Table 1.1.

TABLE 1.1 Simple formulas in probability

(margin note: good summary)

The event A occurs	$P(A)$
The event A does not occur	$P(\bar{A}) = 1 - P(A)$
Either A or B or both occur	$P(A \cup B) = P(A) + P(B) - P(A \cap B)$
A and B occur simultaneously	$P(A \cap B) = P(A\,|\,B)P(B) = P(B\,|\,A)P(A)$
The event A occurs given that B occurs	$P(A\,|\,B) = \dfrac{P(A \cap B)}{P(B)}$

$$= \frac{P(B\,|\,A)\,P(A)}{P(B\,|\,A)\,P(A) + P(B\,|\,\bar{A})\,P(\bar{A})}$$

$$= 1 - P(\bar{A}\,|\,B)$$

PROBLEMS

1. Two dice are tossed. Determine the probability that:
   (a) the sum of the points is 7; (b) the sum is even; (c) the difference is 3.

2. Given two independent events A and B with $P(A) = 0.5$ and $P(B) = 0.8$, determine the probability of
   (a) A and B; (b) A or B; (c) neither A nor B.

3. Given $P(A) = 0.6$, $P(B) = 0.4$, $P(A \cap B) = 0.2$, evaluate
   (a) $P(A \cup B)$; (b) $P(A\,|\,B)$; (c) $P(B\,|\,A)$; (d) are A and B independent?

4. Each of three events, A, B, and C has a probability of 0.4. Event A is mutually exclusive of each of the other two events, but $P(B \cap C) = 0.2$. Determine the probability that at least one of the events will occur.

5. Let A and B be events in a sample space, such that $P(A) = 0.4$, $P(B) = 0.3$, and $P(A$ and $B) = 0.2$. Find the probabilities of the events:
   (a) $\bar{A}$ given B; (b) A or $\bar{B}$; (c) $\bar{A}$ and $\bar{B}$.

6. Let A, B, C be the events of a sale of a computer to three companies, respectively. Suppose that $P(A) = 0.7, P(B) = 0.5, P(C) = 0.6, P(B \cap C) = 0.3$, and the probability of making the sale to the first but not to the second is 0.3. Further, the probability that the first company, but neither the second nor the third, will buy is 0.1. (a) Determine the probability that the first or the second companies will not buy, but the third will purchase the machine. (b) What is the probability that at least one sale will be made? (c) What is the chance that all three companies will buy?

7. Two coins and a die are tossed. What is the probability that both coins fall tails and the die shows six?

8. It is generally accepted that the occurrences of color blindness and deaf-
ness are independent in males. Fill four missing entries in the following
table summarizing the relative frequencies.

	Deaf	Not deaf	Total
Color-blind			8.0%
Not color-blind			92.0%
Total	0.2%	99.8%	100.0%

9. A shipment of electronic tubes contains 96 good and four bad tubes.
Two are drawn at random from the lot, and one of them is tested to be
good. What is the probability that the other is good also?

10. The relative frequencies of the incidence for red-green color blindness
based on a large human population are summarized in the following
table.

	Male	Female	Total
Color-blind	4.22%	0.64%	4.86%
Not color-blind	48.48%	46.66%	95.14%
Total	52.70%	47.30%	100.00%

(a) What is the incidence rate (probability) of color blindness for the male?
(b) What is the estimated probability that a color-blind individual is
a male?

11. The probability that Mr. Doe will vote in an election is 0.7. The probability that both Mr. and Mrs. Doe will vote is 0.63. If you see Mr. Doe
voting at the polls, what is the probability that his wife is also voting?

12. Among the members of a country club, 5% are business executives and
20% of all club members have incomes over $70,000 per year. If it is
known that 80% of the members who are business executives have
incomes over $70,000, determine (a) the probability that a member is
an executive if you are told that the man has an income over $70,000;
and (b) the combined percentage of executives or those making over
$70,000.

13. According to a recent U.N. yearbook, the divorce rate in Sweden, the
United States, and France is 60%, 44%, and 10%, respectively. Consider
three randomly selected women: a Swede, an American, and a French-
woman, all newly married. (a) What is the probability that all three will
be divorced some day? (b) Find the probability that at least one will be
divorced. (c) What is the probability that exactly one will be divorced?

14. It is estimated that in a diagnosis of renovascular hypertension the intra-
venous pyelogram gives a positive indication in 100% of patients with
the disease, but a false positive indication of 10% for those with no

disease. It is further estimated that 10% of a certain population has the disease. (See *New England Journal of Medicine*, 1974, Vol. 291, p. 1115.) Determine the probability that an individual in the population has reno-vascular hypertension if the test shows a positive result.

15. A man leaves his work at 5 P.M. He gets off 80% of the time at 5 P.M., but 20% of the time he leaves 5 minutes before 5 P.M. If he leaves early, he gets home by 6 P.M. 90% of the time; but if he leaves at the regular quitting time, he arrives home by 6 P.M. 20% of the time because of rush-hour traffic. If he gets home after 6 P.M. on one day, what is the prob-ability that he was not able to get off early on that day assuming that he didn't stop for a beer or some other business?

16. A commando operation is to begin at a certain midnight. The probability of rain is estimated as 0.8. If it rains, the probability of successful opera-tion is 0.9, while it is only 0.6 if it does not rain. What is the probability of success?

17. The probability that a friend of yours dated last night is 0.1. The prob-ability that he will be at his 8 A.M. class is 0.2 if he dated and 0.9 if he did not. What is the chance he will show up in class this morning?

18. In Crestwood, 60% of the voters are Republicans and 40% are Demo-crats. There are two candidates for mayor: *A* and *B*. In the election, 70% of the Republicans and 40% of the Democrats voted for *A*, and 30% of the Republicans and 60% of the Democrats voted for *B*. (a) If a voter is chosen at random, what is the probability that he voted for *A*? (b) If an individual said that he voted for *B*, what is the probability that he is a Republican?

19. An electronic assembly consists of two subsystems, *A* and *B*. The follow-ing probabilities are assumed to be known:

$P(A$ alone fails$) = 0.05$, $P(A$ and B fail$) \doteq 0.05$, $P(B$ fails$) = 0.10$.
and B doesn't fail
Evaluate the following probabilities:

(a) $P(B$ fails if A fails$)$ (b) $P(B$ fails alone$)$.

20. Show that if A and B are independent events, so are $\bar{A}$ and B, $\bar{A}$ and $\bar{B}$.

21. Suppose there are 100 heart patients in a certain hospital and it is found that all of the 100 patients are coffee drinkers. Is it rational to state that coffee drinking is associated with heart conditions? Discuss.

22. Everify Eq. (1.11). Hint: $P(A|B) = P(A \cap B)/P(B)$ and $P(B) = P(B|A) P(A) + P(B|\bar{A})P(\bar{A})$.

23. Following the hint given for Prob. 22, verify Eq. (1.12).

24. Show that a diagnostic test is 100% accurate if and only if both sensitivity and specificity are 100%.

2

Description of Random Variables

2.1 Introduction

A *population* from a statistical point of view may be loosely defined as the totality of the elements of any kind that have one or more characteristics or properties in common. Thus, for example, we may speak of the population of patients with a certain disease, the population of a specific electric part produced by a manufacturer, the population of a certain bacteria, the population of all one dollar bills, and so on. The random observation from a population may result in any of a number of possible outcomes about the specified characteristic or property. The random variable (often referred to simply as the variable) is the numerical quantity whose value is determined on the outcome. Therefore, a specified characteristic of a population is completely determined if the characteristic of the variable is known.

There are two general classes of variables, discrete variables and continuous variables. A *discrete variable* is one whose values can take only discrete points. A *continuous variable* is one that can assume any value in a certain range. We shall use capital letters such as X and Y to denote variables that we can observe, and constant numbers by the lower case letters. Thus, for example, $P(X = x)$ denotes the probability that a variable X can take value x, and $P(a < Y < b)$, the probability that a variable Y can assume a value between a and b. This chapter deals with the description and properties of the variable which defines the population characteristic.

2.2 Theoretical Distribution of Variables

Consider the variable X which represents the sum of the points showing on two dice. The X clearly is a discrete variable. The possible values x of the X and the corresponding probabilities are given by

19

x	2	3	4	5	6	7	8	9	10	11	12
$P(X = x)$	1/36	2/36	3/36	4/36	5/36	6/36	5/36	4/36	3/36	2/36	1/36

A table like the above, which gives the probability $P(X = x)$ for each possible value x of a discrete variable X, is called a <u>theoretical distribution</u>, or more precisely a *frequency function*, of the discrete variable X. A frequency function of a variable X is denoted by $p(x)$. In other words,

$$p(x) = P(X = x). \qquad (2.1)$$

The graphical sketch of the frequency function $p(x)$ of the above variable X is shown in Fig. 2.1, where the lengths of the vertical bars represent the magnitude of the probabilities.

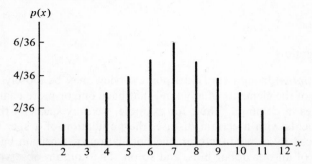

Fig. 2.1   Frequency function of the variable which is the sum of two dice

If $p(x)$ is known, the probability that X can be less than or equal to x is calculated from

$$P(X \leq x) = \sum_{t \leq x} p(t). \qquad (2.2)$$

The probability given by (2.2) is called the *cumulative distribution function* of a variable X and denoted by $F(x)$. <u>Since a variable always assumes one of a number of possible values, the sum of probabilities must be equal to one.</u> To be precise,

$$\sum_{\text{all } x} p(x) = 1.$$

In summary, the probability of a discrete variable is characterized by a frequency function $p(x)$ which satisfies two conditions:

(a) $p(x) \geq 0$ for all x, $\qquad (2.3)$

(b) $\sum_{\text{all } x} p(x) = 1.$ $\qquad (2.4)$

We often say that a variable X has a distribution $p(x)$ if it is the frequency function of X.

For a continuous variable X, there is a function $f(x)$ corresponding to the frequency function $p(x)$ in the case of a discrete variable. The function is called a *density function* of X and is a continuous curve defined over the range of X. The density function is nonnegative for all values of X, and the curve is "normalized" so that the total area under the curve is equal to one. The last mentioned property is necessary, since we want to interpret the area under the curve as a probability. To be precise, $f(x)$ has the property that, given any two values a and b with $a < b$, the probability that a variable X is between a and b is equal to

$$P(a < X < b) = \int_a^b f(x)\, dx. \tag{2.5}$$

An example of a density function is given by

$$f(x) = 2e^{-2x}; \quad x \geq 0.$$

This density function $f(x)$ is shown graphically in Fig. 2.2. It is noted that $P(a < X < b)$ represents the area under the graph between a and b.

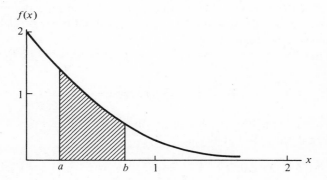

Fig. 2.2   Graph of density function $f(x) = 2e^{-2x}$

The *cumulative distribution function* of a continuous variable with density function $f(x)$ is denoted by $F(x)$ and given by

$$F(x) = \int_{-\infty}^x f(t)\, dt, \tag{2.6}$$

which is analogous to (2.2).

Corresponding to the two conditions (2.3) and (2.4) for the frequency function, a density function satisfies

$$\text{(a)} \quad f(x) \geq 0 \quad \text{for all } x, \tag{2.7}$$

$$\text{(b)} \quad \int_{-\infty}^{\infty} f(x) = 1. \tag{2.8}$$

As in the case of the discrete variable, we shall simply say that a variable X has a distribution $f(x)$ if it is the density function of X.

It is noted that when X is continuous, $P(X = a) = 0$ for any given value a, since the area of the line $X = a$ is zero. In other words, the probability that a continuous variable is equal to any specified value is zero. Therefore, for example, $P(a \leq X \leq b) = P(a < X < b)$ if X is a continuous variable. On the other hand, if X is a discrete variable, we have $P(a \leq X \leq b) \geq P(a < X < b)$.

We should remember that not all variables have the same form of distributions, and that there are many different frequency functions and density functions. The distribution serves as a model for the variables representing certain characteristics of populations. Some important distributions in the application of statistics are described in the next chapter.

Example 2.1 Consider a variable X with the following distribution:

$$f(x) = \frac{2 - x}{2}, \qquad 0 \leq x \leq 2$$
$$= 0, \qquad x < 0 \quad \text{or} \quad x > 2.$$

(a) Show that the $f(x)$ is a density function. (b) Determine the probability that the variable will be between 0.5 and 0.7. (c) What is the probability that the variable will be greater than 1?

(a) First, it is trivial to show that the function satisfies (2.7). Next, we have

$$\int_{-\infty}^{\infty} f(x)\, dx = \int_{0}^{2} \frac{2 - x}{2}\, dx = x - \frac{x^2}{4}\Big|_{0}^{2} = 1$$

satisfying (2.8); therefore, it is a density function.

(b) The desired probability is

$$P(0.5 < X < 0.7) = \int_{0.5}^{0.7} \frac{2 - x}{2}\, dx = x - \frac{x^2}{4}\Big|_{0.5}^{0.7} = 0.14.$$

(c) The probability that the X will be greater than 1 is

$$P(X > 1) = \int_{1}^{2} \frac{2 - x}{2}\, dx = x - \frac{x^2}{4}\Big|_{1}^{2} = 0.25.$$

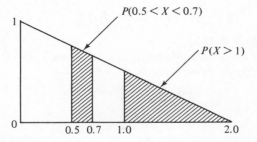

Fig. 2.3 Density function $f(x) = (2 - x)/2$ and regions representing $P(0.5 < X < 0.7)$ and $P(X > 1)$

The graph of the density function and the region representing the probabilities of parts (b) and (c) are illustrated in Fig. 2.3. For this example, it is easy to answer all three questions geometrically without using calculus. ▲

2.3 Simple Description of Theoretical Distributions

It is often desirable to describe the distribution of a variable on the basis of a few of its simple features. Among these, two important features are the measure of central location and that of dispersion or variation.

The most commonly used measures of location are the *mean* and the *median*. The mean of a variable X can be interpreted as a weighted average of the possible values of X, the weight being the probability. To be precise, let X be a discrete variable with possible values $x_1, x_2, \ldots$, and $p(x)$ be the frequency function. Then, the mean μ of the variable is given by

$$\mu = \sum_{i=1}^{\infty} x_i\, p(x_i). \qquad (2.9)$$

If X is a continuous variable with a distribution $f(x)$, the mean is given by

$$\mu = \int_{-\infty}^{\infty} x f(x)\, dx. \qquad (2.10)$$

The mean of X in the distribution represents the long-run average value of the random quantity which X represents. Physically, it is the center of gravity of the distribution. The mean is often referred to as the population mean, or the true mean. It is also called the *expectation*, or the *expected value* of X, and denoted alternatively by $E(X)$.

We shall occasionally use the notation $E(\ )$, which denotes a mathematical operation of "averaging" of a variable or a function of a variable within parentheses. It will mean that the expectation is to be computed for whatever quantity appears in parentheses. Thus, for example, $E(Y^2)$ denotes the expected value of the variable Y^2 and $E(X + Y)$ the expected value of the variable $X + Y$.

The second common measure of location is the median which we shall denote by M. The median is the point at which a vertical line bisects the distribution into two equal parts each containing the probability $\frac{1}{2}$. Let X be a continuous variable with the density function $f(x)$. Then the median M is the point which satisfies

$$\int_{-\infty}^{M} f(x)\, dx = \tfrac{1}{2} \qquad (2.11)$$

Alternatively, if $F(x)$ is the cumulative distribution of X then the median is a root of the equation

$$F(x) = \tfrac{1}{2}. \qquad (2.12)$$

Although the median can be defined for the discrete variable in an analogous manner, it is often difficult to determine, and furthermore, the median is neither unique nor a very precise number in the discrete case. For this reason most applications involving the median deal with continuous variables.

The relative positions of mean μ and median M of a positively skewed continuous distribution, which has a longer tail to the right, are depicted in Fig. 2.4. It is of course possible that the median may be greater than the mean in other types of distributions. As might be expected, the mean and median coincide if a distribution is symmetric.

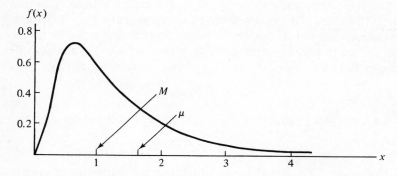

Fig. 2.4 Relative positions of median and mean of a positively skewed distribution

Another less useful measure of location is the *mode* which is roughly a most probable value which X assumes. For a continuous variable a mode is the point, if it exists, at which the density $f(x)$ is a maximum.

It is reasonable to assume that variation of a variable represents its spread about a measure of location. The single most important such measure of the variation is given by the variance or its positive square root known as the standard deviation. The *variance* of X denoted by σ^2 is the expectation of a quantity defined by $(X - \mu)^2$. If most of the distribution is concentrated around the mean the variance will be small; in particular, it will be zero if the distribution is concentrated at one point. Mathematically, the variance of a discrete variable X is defined as

$$\sigma^2 = \sum_{i=1} (x_i - \mu)^2 p(x_i), \tag{2.13}$$

or equivalently,

$$\sigma^2 = \sum_{i=1} x_i^2 p(x_i) - \mu^2. \tag{2.14}$$

For a continuous variable, the variance of X is given by

$$\sigma^2 = \int (x - \mu)^2 f(x)\, dx, \tag{2.15}$$

which is algebraically equivalent to

$$\sigma^2 = \int x^2 f(x)\, dx - \mu^2. \tag{2.16}$$

The variance is a weighted average of the squared deviation of the possible values of X from the mean, the weight being the probabilities.

The positive square root of the variance is called the *standard deviation*, and is denoted by σ. From (2.13) and (2.16), we can see that the larger the variation in the distribution, the larger is the standard deviation σ.

Example 2.2 Suppose that a fair die is tossed, and the variable X designates the number of points showing. The possible values of X are 1, 2, ..., 6, and since it is a fair die, the frequency function is given by $p(1) = p(2) = \cdots = p(6) = \frac{1}{6}$. The mean is

$$\mu = 1(\tfrac{1}{6}) + 2(\tfrac{1}{6}) + \cdots + 6(\tfrac{1}{6}) = 3.5.$$

To determine the median, we see that no single value satisfies (2.11), since

$$\sum_{i=1}^{3} p(x_i) = \tfrac{1}{2}, \qquad \sum_{i=3}^{6} p(x_i) = \tfrac{2}{3},$$

and

$$\sum_{i=1}^{4} p(x_i) = \tfrac{2}{3}, \qquad \sum_{i=4}^{6} p(x_i) = \tfrac{1}{2}.$$

In this circumstance, we take the mean of 3 and 4, namely, 3.5, as the median. Using (2.14), the variance of X is computed as

$$\sigma^2 = 1(\tfrac{1}{6}) + 4(\tfrac{1}{6}) + \cdots + 36(\tfrac{1}{6}) - 3.5^2 = 2.92,$$

and the standard deviation is $\sigma = 1.71$. ▲

One interesting fact illustrated by this example is that the mean or median does not necessarily have to be a possible value of X.

Example 2.3 Suppose that a continuous variable X has a density function

$$f(x) = 2e^{-2x},\ x \geq 0.$$

First, it can be shown that $f(x)$ is a density function. The mean of X is

$$\mu = \int_0^\infty x2e^{-2x}\, dx = \frac{e^{-2x}}{2}(-2x - 1)\Big|_0^\infty = \frac{1}{2}.$$

To determine the median, we let

$$\int_0^M 2e^{-2x}\, dx = -e^{-2x}\Big|_0^M = 1 - e^{-2M} = \frac{1}{2}.$$

Solving the above equation for M yields $M \approx 0.35$. Thus, the median is smaller than the mean in this particular example. The variance of X is

$$\sigma^2 = \int_0^\infty x^2 2e^{-2x}\, dx - (\tfrac{1}{2})^2 = \tfrac{1}{4}. \quad ▲$$

*2.4 More about Mean and Variance of Variables

The following formulas regarding the mean or expectation are worthwhile noting. Each of the three formulas can be easily proved.

(a) The mean of a variable aX, where a is a constant and X is a variable, is

$$E(aX) = aE(X). \qquad (2.17)$$

(b) The mean of $a + X$ is

$$E(a + X) = a + E(X). \qquad (2.18)$$

(c) Let $X_1, X_2, \ldots, X_n$ be n variables not necessarily independent. Then,

$$E(X_1 + X_2 + \cdots + X_n) = E(X_1) + E(X_2) + \cdots + E(X_n). \qquad (2.19)$$

If $X_1, X_2, \ldots, X_n$ have the common mean μ, then it follows that

$$E(X_1 + X_2 + \cdots + X_n) = n\mu. \qquad (2.20)$$

Let $\bar{X}$ represent the average based on n variables, each of them with the same mean μ. Using (2.17) and (2.20) we obtain a very important equality:

$$E(\bar{X}) = E\left(\frac{\sum\limits_{i}^{n} X_i}{n}\right) = \mu. \qquad (2.21)$$

Similar to the three formulas for the mean, we have the corresponding formulas for the variance of some simple function of a variable.

(a) The variance of a variable $a + X$, where a is a constant, is

$$\text{Var}(a + X) = \text{Var}(X). \qquad (2.22)$$

(b) The variance of aX is

$$\text{Var}(aX) = a^2 \text{Var}(X). \qquad (2.23)$$

(c) Let $X_1, X_2, \ldots, X_n$ be n independent variables. Then,

$$\text{Var}(X_1 + X_2 + \cdots + X_n) = \text{Var}(X_1) + \text{Var}(X_2) + \cdots + \text{Var}(X_n). \qquad (2.24)$$

Thus, if $X_1, X_2, \ldots, X_n$ are independent and have the common variance $\text{Var}(X) = \sigma^2$, then

$$\text{Var}(X_1 + X_2 + \cdots + X_n) = n\sigma^2. \qquad (2.25)$$

Further, if $\bar{X}$ denotes the average of n independent variables, each of them with the same variance σ^2, it follows that

$$\text{Var}(\bar{X}) = \text{Var}\left(\frac{\sum\limits_{i}^{n} X_i}{n}\right) = \frac{n\sigma^2}{n^2} = \frac{\sigma^2}{n}. \qquad (2.26)$$

The results given by (2.21) and (2.26) are particularly important, and should be noted and remembered. In particular, the results may be stated in the following theorem.

Theorem 2A

If $\bar{X}$ denotes the mean of n independent observations from a population with mean μ and variance σ^2, then

$$E(\bar{X}) = \mu,$$

$$\text{Var}\,(\bar{X}) = \frac{\sigma^2}{n}.$$

The theorem implies that the distribution of $\bar{X}$ has the same mean as the population from which the observation is drawn, and that as n increases, the distribution of $\bar{X}$ becomes more concentrated about its mean μ.

Example 2.4 In a game in which a fair coin is tossed, a man wins a dollar if the outcome is a head, and loses a dollar if the outcome is a tail. Suppose that he tosses once and quits if he wins, but tries one more time if he loses on the first bet. What is his expected winning? What is the variance of the expected winning?

He can win one dollar with the probability $\frac{1}{2}$; he can lose two dollars or come out even with both probabilities equal to $\frac{1}{4}$. The amount of winnings denoted by X is a random variable, and the mean value is

$$E(X) = 1(\tfrac{1}{2}) + (-2)(\tfrac{1}{4}) + 0(\tfrac{1}{4}) = 0.$$

Thus, the expected return is 0 dollars, and on the average he would come out even. The variance of X, using (2.14), is given by

$$\text{Var}\,(X) = 1(\tfrac{1}{2}) + (-2)^2(\tfrac{1}{4}) + 0(\tfrac{1}{4}) - 0 = 1.5. \quad \blacktriangle$$

Example 2.5 A man bets \$15 for a game which pays him \$80 if he gets three heads by tossing a coin three times. Is it a fair game in which the man can expect to break even? Suppose that he plays the game ten times. What is the expected value and the standard deviation of the total return?

The probability of getting three heads is $\frac{1}{8}$ using the multiplication theorem on independent events. Let X be the amount of return. Then,

$$E(X) = (-15)(\tfrac{7}{8}) + (65)(\tfrac{1}{8}) = -5.$$

Hence, it is an unfair game. The variance of X is given by

$$\text{Var}\,(X) = [(-15)^2(\tfrac{7}{8}) + (65)^2(\tfrac{1}{8})] - (-5)^2 = 700.$$

If he plays the same game ten times, the expectation and standard deviation of the total return is, using (2.17) and (2.25), \$(−50.0) and \$83.7, respectively. The result may be interpreted as follows: If the game is played repeatedly, he can win \$65 once in a while, but he is more often likely to lose \$15. On the average, he is expected to lose \$5 per game; if he plays ten times, the total loss, on the average, will amount to \$50. $\blacktriangle$

2.5 Summary

This chapter deals with the definitions and concepts related to theoretical distribution of a random variable X. Some important concepts discussed are summarized below.

(a) For a discrete variable X with a frequency function $p(x)$ with possible values $x_1, x_2, \ldots$, the mean and variance of X are

$$\mu = \sum_{i=1}^{\infty} x_i p(x_i), \qquad \sigma^2 = \sum_{i=1}^{\infty} (x_i - \mu)^2 p(x_i).$$

(b) For a continuous variable X with a density function $f(x)$, the mean and variance of X are given by

$$\mu = \int_{-\infty}^{\infty} x f(x)\, dx, \qquad \sigma^2 = \int_{-\infty}^{\infty} (x - \mu)^2 f(x)\, dx.$$

(c) If $\bar{X}$ denotes the mean of a sample of n independent observations from a distribution with mean μ and variance σ^2, then

$$E(\bar{X}) = \mu, \qquad \text{Var}(\bar{X}) = \frac{\sigma^2}{n}.$$

PROBLEMS

let $f(x) = p(x)$

1. Verify that each of the following functions is a frequency function and sketch its graph.

(a) $f(x) = \frac{1}{4}$ for $x = 0$
    $p(x) = \frac{3}{4}$ for $x = 1$
    $= 0$ otherwise.

(b) $f(x) = \frac{1}{3}(\frac{2}{3})^{x-1}$ $x = 1, 2, 3, \ldots$
    $p(x) = 0$ otherwise

(c) $f(x) = \binom{3}{x}\left(\frac{1}{3}\right)^{x}\left(\frac{2}{3}\right)^{3-x}$ for $x = 0, 1, 2, 3$
    $p(x) = 0$ otherwise.

2. Verify that each of the following functions is a density function and sketch its graph.

(a) $f(x) = 2x$ for $0 < x < 1$
    $= 0$ elsewhere.

(b) $f(x) = e^{-x}$ for $x \geq 0$
    $= 0$ for $x < 0$.

(c) $f(x) = \frac{1}{2\sqrt{x}}$ for $0 < x < 1$
    $= 0$ elsewhere.

3. Determine the probability $P(0.5 < X < 2)$ for each of the distributions given in Prob. 1.

4. Find the probability $P(1 \leq X \leq 3)$ for each of the distributions given in Prob. 2.

5. Calculate the mean and variance for each of the distributions given in Prob. 1.

6. Calculate the mean, median, and variance for each of the distributions given in Prob. 2.   a) $2/3$, $\frac{\sqrt{2}}{2}$, $\frac{1}{18}$ b) 1, $\ln 2$, 1 c) $1/3$, $1/4$, $4/45$

7. The following represents the distribution of X which represents the daily demand in units of a certain product. Find the mean and standard deviation of the variable.

satisfy: 1) $0 \le p(x) \le 1$ ok

2) $\sum\limits_{all\,x} p(x) = 1$ ok

x	1	2	3	4	5
$p(x)$	0.1	0.1	0.2	0.4	0.2

mean $= 3.5$
variance $= 1.45$

8. A shop has two machines, A and B, being run independently. The following table gives the distribution of the number of breakdowns for each machine per day.
   (a) Calculate the probability that the total number of breakdowns is less than three.
   (b) Compare the mean number of breakdowns for the two machines.
   (c) Compare the variance of the number of breakdowns for A and B.   1.5676
   (d) What can you conclude from (b) and (c) with regard to the dependability of A and B?   2.31

	Number of breakdowns					
	0	1	2	3	4	5
A	0.15	0.30	0.25	0.20	0.08	0.02
B	0.30	0.20	0.20	0.15	0.10	0.05

9. Suppose that X is a continuous variable with distribution
$$f(x) = 1 + x, \qquad -1 \le x \le 0$$
$$= 1 - x, \qquad 0 \le x \le 1.$$

Calculate the mean, median, and standard deviation of the variable.   0   0 $1/6$

*10. A lottery has one prize of $1000, two prizes of $500, and five of $100. Ten thousand tickets are sold. If the price of a ticket is $.50, is it a good buy?

*11. Assume that 10% of the parcels sent to a certain foreign country do not reach their destination. Suppose that we send two gifts either in a single package or in two separate packages. (a) Compare the probabilities that both gifts will reach their destination by the two methods. (b) What is the probability that at least one will be delivered by each of the two methods? (c) If the gifts cost $150 and $200, respectively, determine the value that can be expected to be delivered by each method.

*12. If a variable is multiplied by a negative constant, find the mean and standard deviation of the resulting variable.

*13. A manufacturer offers the following deal to buyers. A buyer inspects four items from a lot of large items. If he finds no defective item, he pays $10,000 for the lot. If he finds one or more defectives, he pays $8000 for the lot. (a) If the actual proportion of defectives is 5%, what is the expected amount the manufacturer can make per lot? (b) What has to be the quality of the lot if the manufacturer expects to make $9000 per lot?

*14. Let X and Y be two independent variables. For two constants, a and b, prove the following formulas:
  (a) $E(aX + bY) = aE(X) + bE(Y)$
  (b) $\text{Var}(aX + bY) = a^2 \text{Var}(X) + b^2 \text{Var}(Y)$.

3

Some Important
Theoretical Distributions

3.1 Introduction

As observed in Chap. 2, the distribution serves as a probability model for the variables representing real phenomena or certain characteristics of populations. A distribution, or to be precise, a theoretical distribution of a variable X, can be specified by giving its cumulative distribution function. An equivalent and more common way is to give its frequency function or density function depending on whether X is discrete or continuous. All three, the cumulative distribution function, frequency function, and density function, are usually referred to as distributions.

The theoretical distribution of a random variable X has a specific functional form with one or more *parameters*. In order to describe the parameter, consider a lot consisting of n units of a certain item. Suppose that each unit is classified as defective or nondefective. A unit is taken at random from a lot, and the value $X = 0$ is assigned if it is nondefective, and $X = 1$ if it is defective. In this example, of course, X is a random variable, which in this case takes the value 0 or 1. Denote the probability of a defective unit by p, which is given by $P(X = 1)$. We can summarize the probability as follows:

$$P(X = 0) = 1 - p, \qquad P(X = 1) = p.$$

The quantity p is a parameter of the distribution, and its value is usually unknown. Note that a numerical characteristic of the variable or the corresponding population, such as its mean or standard deviation, can be calculated if p is known. From (2.9) and (2.14), we obtain

$$\mu = p, \qquad \sigma = \sqrt{p(1 - p)}.$$

The variable with the above probability distribution is discussed further in Sec. 3.2.

As a second example, suppose we are conducting a study of animal behavior in which blindfolded animals are released at a given point. It is expected that the animals will run randomly in a 360° range. Put more precisely, if X denotes the angle from a certain direction, its density function will be constant over the interval (0°, 360°).

To be more general, consider a continuous variable X whose value can lie only in a certain finite interval (a, b). If the density function is constant over the interval, it can be defined as

$$f(x) = \begin{cases} \dfrac{1}{b-a} & \text{if } a < x < b \\ 0 & \text{otherwise.} \end{cases}$$

The quantities a and b which determine the distribution are the parameters. If a and b are known, we can determine precisely any property of the population characterized by the above distribution. For example, the mean and variance can be calculated from (2.10) and (2.16) as follows:

$$\mu = \frac{a+b}{2}, \qquad \sigma^2 = \frac{(b-a)^2}{12}.$$

The variable with the above density function $f(x)$ is said to have a *uniform*, or *rectangular*, *distribution*. Some readers may be familiar with this distribution because of the computer program known as the random number generator which generates random numbers distributed according to the uniform distribution with $a = 0$ and $b = 1$.

It is worth stating that the parameters of a distribution are not necessarily its mean and standard deviation as illustrated by these examples. The mean and standard deviation, however, are always functions of the parameters, and are often, in a loose sense, referred to as parameters.

As has been stated already, a frequency function or density function can be conceived as a probability model for describing the characteristic of a population. Naturally, we could imagine an infinite number of distributions depending on the population involved. However, we want to select one of the well-known forms to facilitate the analysis, and this is satisfactory for most applications. The important distributions in applications are now discussed.

3.2 Bernoulli Distribution and Binomial Distribution

Consider an experiment which can have only two possible outcomes: conveniently defined as "success" or "failure." For example, a toss of a coin will show either a head or a tail; a patient undergoing a treatment may live or die; a missile may hit or miss a target. Such experiments are called *Bernoulli trials*. Let p be the probability of success and q be the probability of failure, such that $p + q = 1$. The random variable X can be defined conveniently as

$$X = \begin{cases} 1 \text{ if outcome is success} \\ 0 \text{ if outcome is failure.} \end{cases}$$

The random variable X is said to have a *Bernoulli distribution*, and its frequency function $p(x)$ is given by

$$p(x) = \begin{cases} p & \text{if} & x = 1 \\ q & \text{if} & x = 0. \end{cases} \tag{3.1}$$

Before discussing the next distribution, we shall define a mathematical symbol which is used next and in other parts of this text. It is the so-called *binomial coefficient* $\begin{pmatrix} n \\ x \end{pmatrix}$ defined as

$$\begin{pmatrix} n \\ x \end{pmatrix} = \frac{n!}{x!(n-x)!}$$

where, for example, $n!$ denotes n factorial which is given by

$0! = 1$

$$n! = n(n-1)(n-2) \cdots (3)(2)(1).$$

Thus, we have $1! = 1$, $2! = 2$, $3! = 6$, $4! = 24$, etc. $0!$ is defined to be 1.

Now, consider n independent Bernoulli trials and let X be the number of successes out of n trials. The random variable X is said to have a *binomial distribution* and its frequency function defined by

$$p(x) = \begin{pmatrix} n \\ x \end{pmatrix} p^x q^{n-x}, \qquad x = 0, 1, 2, \ldots, n, \tag{3.2}$$

where $q = 1 - p$, gives the probability of exactly x successes in n trials when the probability of each success is p. The mean and variance of the binomial variable are given by

$$\mu = np, \qquad \sigma^2 = npq. \tag{3.3}$$

b) mean # of def bolts when 3 bolts drawn from lg lot w/ 5% def
n= 3 p= .05
u= np

It is evident that a Bernoulli distribution can be regarded as a special case of a binomial distribution with $n = 1$. The graphical sketch of a binomial distribution when $n = 5$ and $p = 0.6$ is shown in Fig. 3.1.

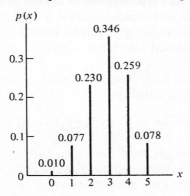

Fig. 3.1   Binomial distribution for $n = 5$ and $p = 0.6$

Example 3.1 If, in general, 30% of patients afflicted with a certain disease die from it, (a) what is the probability that none will die in a group of five? (b) What is the probability at least three will die? The desired probabilities are determined as

$$P(X = 0) = p(0) = \frac{5!}{0!5!}(0.3)^0(0.7)^5 = 0.168$$

and

$$P(X \geq 3) = \sum_{x=3}^{5} \frac{5!}{x!(5 - x)!}(0.3)^x(0.7)^{5-x} = 0.163. \quad \blacktriangle$$

3.3 Geometric Distribution

 (a) toss die till get a one

Let X denote the number of trials required to obtain the first "success" in a sequence of independent repeated Bernoulli trials in which the probability of success at each trial is p. Thus, X assumes the possible values $1, 2, \ldots$. Since $X = x$ if and only if the first $x - 1$ trials result in failure and the next trial results in success, the frequency function of X is given by

$$p(x) = pq^{x-1}, \qquad x = 1, 2, \ldots$$

where $q = 1 - p$. The random variable X characterized by this frequency function is said to have a *geometric distribution* with parameter p. The mean and variance of the random variable with the geometric distribution are given by

$$\mu = \frac{1}{p}, \qquad \sigma^2 = \frac{1-p}{p^2}.$$

Example 3.2 Suppose the cost of conducting an experiment is $500. If the experiment fails, an additional cost of $60 is incurred to clean up before the next trial. Assume that the probability of success on any given experiment is 0.2 and the individual trials which require one week each are independent. (a) What is the probability that less than four weeks are required to obtain the first successful experiment? (b) What is the total expected cost in obtaining one successful result?

Let X be the number of weeks to achieve the first success. The first question is answered by calculating

$$P(X < 4) = \sum_{x=1}^{3} (0.2)(0.8)^{x-1} = 0.49.$$

For the second problem, let C denote the total cost and X the number of experiments to achieve the desired result. We have $C = 500X + 60(X - 1) = 560X - 60$. Using $E(X) = \mu = 1/p$ along with (2.17) and (2.18), we calculate

$$E(C) = E(560X - 60) = 560E(X) - 60$$

$$= 560\frac{1}{0.2} - 60 = 2740$$

Therefore, \$2740 is the expected cost of the experiment. ▲

3.4 Poisson Distribution

There are essentially two reasons that the Poisson distribution has an important role in many applications. The first is due to the relation between the Poisson distribution and the binomial distribution. If the number of Bernoulli trials n is very large, the computations with (3.2) are rather prohibitive. Suppose that, in addition to n being large, p is very small so that the product

$$\lambda = np$$

is of moderate magnitude. In this circumstance, the binomial frequency function $p(x)$ is very well approximated by the *Poisson frequency function* defined by

$$p(x) = \frac{e^{-\lambda}\lambda^x}{x!}, \qquad x = 0, 1, 2, \ldots. \qquad (3.4) \quad \longleftarrow$$

The accuracy of the approximation increases as n increases for a given p or as p decreases for a fixed n, or as n increases and p decreases. Both the mean and the variance of the Poisson random variable with the frequency function (3.4) are equal to λ. That is,

$$\mu = \lambda, \qquad \sigma^2 = \lambda. \qquad (3.5) \quad \longleftarrow$$

A Poisson distribution with $\lambda = 1.5$ is sketched in Fig. 3.2.

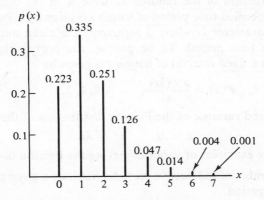

Fig. 3.2 Poisson distribution with $\lambda = 1.5$

Example 3.3 Suppose the probability of an electronic tube being defective is 0.01 and that a random sample of $n = 100$ such tubes is obtained. What is the probability that the sample contains two or less defective tubes? Using the Poisson approximation with $\lambda = np = 1$ the probability is

$$P(X \le 2) = \sum_{x=0}^{2} \frac{e^{-1}}{x!} = 0.9197.$$

Table 3.1 compares the probability of exactly x successes (or that of defective

TABLE 3.1 Poisson approximation when $n = 100$ and $p = 0.01$

X	0	1	2	3	4	5	6	7
Binomial	0.3660	0.3697	0.1849	0.0610	0.0149	0.0029	0.0005	0.0001
Poisson	0.3679	0.3679	0.1839	0.0613	0.0153	0.0031	0.0005	0.0001

tubes in this example) computed by (3.2) and (3.4). If we used the binomial frequency function in this example the desired probability would have been 0.9206. ▲

The Poisson distribution, besides providing an approximation to the binomial distribution, plays an important role in its own right. It can be used to represent and model many observational phenomena.

Suppose that the random phenomena satisfy the following three conditions:

1. Events occurring in one time interval are independent of those occurring in any other mutually exclusive interval.
2. The probability of events occurring is proportional to the length of the time interval.
3. The probability of two or more events occurring in an infinitesimally small interval is negligible.

Then, the distribution of the random variable X of the number of events occurring in a specified time period of length t is given by a Poisson distribution with the parameter λt where λ represents the mean number of occurrences per unit time period. To be precise, the probability of exactly x occurrence(s) in a fixed interval of length t is given by

$$P(x) = \frac{e^{-\lambda t}(\lambda t)^x}{x!}, \quad x = 0, 1, 2, \ldots. \tag{3.6}$$

The mean and variance of the Poisson distribution of the form (3.6) are

$$\mu = \lambda t, \qquad \sigma^2 = \lambda t. \tag{3.7}$$

Conceivable examples of variables having the Poisson distribution are:

(a) The number of radioactive particles reaching a given position during a time period.
(b) The number of telephone calls arriving at a switchboard during a specified time period.

(c) The number of heart attacks in a given period of time in a city.

(d) The number of cells that go through exactly k chromosome interchanges in a given period of time.

(e) The number of tornadoes observed in a fixed time period in a state.

Example 3.4 Workers in a certain factory incur accidents at the rate of two per week. (a) What is the probability that there will be three or less accidents during the next three weeks? (b) Calculate the probability that there will be one or no accident in each of the next three weeks.

It is given that $\lambda = 2$. Let X denote the number of accidents in the next three weeks. The answer to the first question is

$$P(X \leq 3) = \frac{e^{-6}6^0}{0!} + \frac{e^{-6}6}{1!} + \frac{e^{-6}6^2}{2!} + \frac{e^{-6}6^3}{3!} = 0.151.$$

To solve the second problem, let Y denote the number of accidents in a week. Then

$$P(Y \leq 1) = \frac{e^{-2}}{0!} + \frac{e^{-2}2}{1!} = 0.406.$$

Using the multiplication theorem for three independent random variables, we have for the second answer

$$[P(Y \leq 1)]^3 = 0.067.$$

The reader may wish to think why the probability computed for the second question is smaller than that for the first question. ▲

3.5 Exponential Distribution

A continuous random variable X assuming all nonnegative values is said to have an *exponential distribution* with parameter $\lambda > 0$ if its density function is given by

$$f(x) = \begin{cases} \lambda e^{-\lambda x}, & x > 0 \\ 0, & \text{elsewhere.} \end{cases}$$

The shape of an exponential density function is sketched in Fig. 3.3.

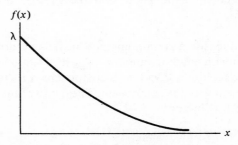

Fig. 3.3   Exponential distribution $f(x) = \lambda e^{-\lambda x}$

The cumulative distribution function of an exponentially distributed variable calculated from (2.6) is given by

$$F(x) = \begin{cases} 1 - e^{-\lambda x}, & \text{for } x \geq 0 \\ 0, & \text{for } x < 0. \end{cases}$$

The mean and variance of the exponential distribution are

$$\mu = \frac{1}{\lambda}, \quad \sigma^2 = \frac{1}{\lambda^2}. \tag{3.8}$$

The exponential distribution is used often as the model for describing lifetime data, for example, the life of electronic components and survival times of experimental animals. It is also used in describing the time intervals between accidents or the time intervals between breakdowns of a component of a system. In this regard the parameter λ is often called the *failure rate* of the exponential distribution.

A random variable X with an exponential distribution is characterized by the following properties: (a) Consider for s, $t > 0$, $P(X > s + t \mid X > s)$;

$$P(X > s + t \mid X > s) = \frac{P(X > s + t)}{P(X > s)} = \frac{e^{-\lambda(s+t)}}{e^{-\lambda s}} = e^{-\lambda t}.$$

Hence

$$P(X > s + t \mid X > s) = P(X > t). \tag{3.9}$$

Suppose that X denotes the lifetime of a component. Then, in words, (3.9) says that if a component lasted at least s time units, the probability of its lasting t or more additional time units is the same as the probability of the component lasting t or more time units independent of the fact that it has lasted at least s time units already. This is an inherent property of the time prior to failure of a component if it is not subject to wear or fatigue. (b) Suppose that a certain event occurs according to the Poisson distribution with the mean rate λ. Then, the distribution of the time intervals between successive occurrences of the event is given by an exponential distribution with the same parameter λ. The converse is also true, and there is a one-to-one relation between the exponential and the Poisson distribution of the form given by (3.6).

Example 3.5 Suppose a certain component fails according to an exponential distribution with a failure rate $\lambda = \frac{1}{1000}$ hr. Determine a time interval h such that the probability is 0.95 that a component will last at least h.

Let T be the lifetime of the component. The density function of T is given by an exponential distribution:

$$f(t) = \frac{1}{1000}e^{-(1/1000)t}, \quad t > 0.$$

The probability of a component lasting at least h hours is

$$P(T \geq h) = \int_h^\infty \tfrac{1}{1000} e^{-(1/1000)t}\, dt$$
$$= e^{-0.001h}$$

Solving the equation $e^{-0.001h} = 0.95$, we obtain $h = 51.3$. Thus, the answer to the problem is 51.3 hr, or about 51 hr. ▲

3.6 Normal Distribution

A continuous random variable X is said to have a *normal distribution*, or *Gaussian distribution*, if its density function has the form

$$f(x) = \frac{1}{\sqrt{2\pi}\sigma} e^{(-1/2)[(x-\mu)/\sigma]^2}, \quad -\infty < x < \infty. \tag{3.10}$$

We shall frequently use the notation: X has $N(\mu, \sigma^2)$ if its density function is given by (3.10). The graph of normal density function with $\mu = 1$ and three different values of σ is shown in Fig. 3.4.

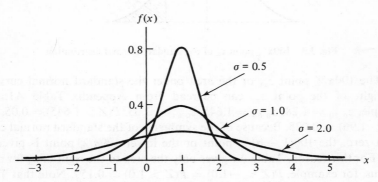

Fig. 3.4 Normal density functions with $\mu = 1$ and $\sigma = 0.5, 1.0,$ and 2.0.

The normal distribution possesses a symmetrical bell-shaped curve (although by no means are all bell-shaped curves represented by the normal distribution). There are two parameters, μ and σ, which determine the central position and the extent of spread of the curve. In fact, these two parameters are the mean and standard deviation of the variable X. For every normal curve, the area between $\mu \pm k\sigma$ where k is any fixed number is constant. For example,

   (a) $\mu \pm \sigma$ contains 68.26% of the area under the curve,
   (b) $\mu \pm 2\sigma$ contains 95.44%, and
   (c) $\mu \pm 3\sigma$ contains 99.74%.

A normal random variable with $\mu = 0$ and $\sigma = 1$ is called a standard normal variable and is commonly denoted by Z: that is, Z has $N(0, 1)$. Its density function, called the standard normal density function, is given by

$$f(z) = \frac{1}{\sqrt{2\pi}} e^{-z^2/2}.$$

The upper $100\alpha\%$ point or, more simply, the $100\alpha\%$ point, $(0 \leq \alpha \leq 1)$, of the standard normal curve is denoted by z_α. In other words, z_α is the value which satisfies

$$P(Z > z_\alpha) = \alpha.$$

Fig. 3.5 illustrates the point z_α for given α.

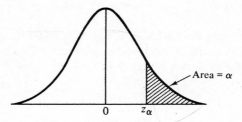

Fig. 3.5 $100\alpha\%$ point z_α of the standard normal distribution

The $100\alpha\%$ point z_α, or the area under the standard normal curve to the right of the point z_α, can be read from Appendix Table A1. For example, $z_{0.10} = 1.28$, $z_{0.05} = 1.645$, $z_{0.01} = 2.33$, $P(Z > 1.645) = 0.05$, and $P(Z > 1.96) = 0.025$. Because of the symmetry of the standard normal curve about zero, the $100(1 - \alpha)\%$ point or the lower $100\alpha\%$ point is given by $-z_\alpha$. For this reason, Table A1 gives only the area to the right of z_α for $z_\alpha \geq 0$. Thus, for example, $P(Z < -1.0) = P(Z > 1.0) = 0.159$. Note that Table A1 can be used to find the area under any part of the standard normal curve. For example, $P(-1 < Z < 2) = 0.818$; the process of obtaining this probability is illustrated in Fig. 3.6.

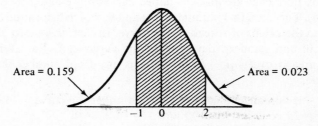

Fig. 3.6   Area under -1 to 2 for the standard normal distribution

Any normal variable X can be transformed into the standard normal variable Z by the transformation

$$Z = \frac{X - \mu}{\sigma}. \tag{3.11}$$

This formula enables us to find the point z on the standard normal curve that corresponds to any point x on a nonstandard normal curve. And this relation enables us to use Table A1 to find the area under any part of any nonstandard normal curve. To be precise, suppose it is desired to obtain the probability $P(a < X < b)$ where X is a normal random variable with mean μ and variance σ^2. To do this note that

$$P(a < X < b) = P\left(\frac{a - \mu}{\sigma} < \frac{X - \mu}{\sigma} < \frac{b - \mu}{\sigma}\right)$$

$$= P\left(\frac{a - \mu}{\sigma} < Z < \frac{b - \mu}{\sigma}\right). \tag{3.12}$$

Thus, the area between a and b for X is given by the area between $(a - \mu)/\sigma$ and $(b - \mu)/\sigma$ of the standard normal curve. Just one table is needed for the entire class of normal distributions!

Example 3.6 Suppose that the mean stature of college girls is 5 ft 6 in. and the standard deviation, 2.4 in. Assuming that the height X has a normal distribution, determine the proportion of girls with statures less than 5 ft 2 in., and that of girls between 5 ft 3 in. and 5 ft 8 in. in height.

Using the inch as the basic unit, the desired probabilities are computed as:

$$P(X < 62) = P\left(Z < \frac{62 - 66}{2.4}\right) = P(Z < -1.67) = 0.048,$$

and

$$P(63 < X < 68) = P\left(\frac{63 - 66}{2.4} < Z < \frac{68 - 66}{2.4}\right)$$

$$= P(-1.25 < Z < 0.83) = 0.691. \quad \blacktriangle$$

The normal distribution is important because many random phenomena can be approximately described by the distribution. Furthermore, the normal distribution can be used to approximate other distributions.

As one important application, consider the number of successes X in n independent Bernoulli trials. Recall that when n is large and p is small, the Poisson distribution is used to approximate the frequency function of X. Suppose that n is large but p is not too small. The following theorem is relevant.

Theorem 3A

For large n the following random variable

$$Z = \frac{X - np}{\sqrt{npq}} \qquad (3.13)$$

has an approximate standard normal distribution.

It follows from Eq. (3.3) that the mean and variance of the variable Z defined in Eq. (3.13) are 0 and 1, respectively. The amazing fact is that Z approaches a normal distribution as n becomes large. Theorem 3A states that X is normally distributed with the mean np and variance npq if the sample size n is large.

The accuracy of the approximation increases as p approaches 0.5 for a fixed n or as n increases for a fixed p, or both happen simultaneously. In particular, the probability $P(a \leq X \leq b)$ when n is moderately large, and p is not small or not too large, or irrespective of p when n is large, is fairly well approximated by

$$P\left(\frac{a - np - 0.5}{\sqrt{npq}} < Z < \frac{b - np + 0.5}{\sqrt{npq}}\right), \qquad (3.14)$$

where Z has the standard normal distribution.

Example 3.7 In promoting a certain medicine, a TV commercial claimed that 30% of the population suffers from a certain symptom. Suppose that in a random sample of size $n = 100$, only 14 suffered from the symptom. Is the TV commerical justified?

The probability that 14 or less individuals out of $n = 100$ suffer from the symptom, assuming that $p = 0.3$, is

$$P(0 \leq X \leq 14) = P\left(\frac{-30 - 0.5}{\sqrt{21}} \leq Z \leq \frac{14 - 30 + 0.5}{\sqrt{21}}\right) = 0.0003.$$

The probability is very small, and the claim of 30% is very difficult to believe. The exact probability using (3.2) gives 0.0002. ▲

For $n = 100$ and $p = 0.3$ the nicety of the normal approximation (3.14) can be judged from Table 3.2 for various values of a and b.

TABLE 3.2 Normal approximation when $n = 100$ and $p = 0.3$

a	b	Binomial	Normal
0	14	0.0002	0.0003
15	20	0.0163	0.0188
21	26	0.2089	0.2034
27	29	0.2379	0.2341
31	36	0.3710	0.3785
46	48	0.0005	0.0003

Table 3.2 shows that the normal approximation of a binomial distribution is satisfactory if n is large. To be more precise, the error of the normal approximation will be small if npq is large. On the other hand, as we recall, when n is large and p is small, we use the Poisson approximation. This suggests that for large values of n when p is small, we can approximate a binomial distribution with either the normal or the Poisson distributions. This in turn suggests that for large values of n the normal distribution can be used to approximate the Poisson distribution. Suppose that a random variable X has a Poisson distribution with mean λ. If λ is large, say $\lambda > 10$, the variable

$$Z = \frac{X - \lambda}{\sqrt{\lambda}}$$

has an approximate standard normal distribution. The mean and variance of Z are 0 and 1, respectively, as in the case of Z defined by (3.13). The probability $P(a < X < b)$ can be calculated using the standard normal approximation:

$$P\left(\frac{a - \lambda}{\sqrt{\lambda}} < Z < \frac{b - \lambda}{\sqrt{\lambda}}\right).$$

The relations between the various approximations are summarized in Fig. 3.7.

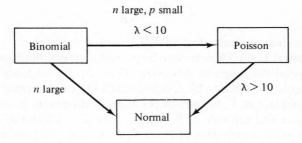

Fig. 3.7 Relations between various approximations

The rationale that enables us to use the normal distribution to approximate others when the sample size is large is not just restricted to the binomial and the Poisson distributions, but extends widely. The underlying principle will be examined and discussed next.

Theorem 3A represents only a special case of a very important general result. In order to realize this, recall that the binomial random variable X can be represented as the sum of the Bernoulli variables. That is, $X = X_1 + X_2 + \cdots + X_n$ where $X_i = 1$ if the ith trial is a success and $X_i = 0$ if it is a failure. Also recall that the mean and variance of X is np and npq, respectively.

If any random variable X may be represented as a sum of any n independent random variables, then in general the sum X, for large n, is approximately normally distributed. This result is known as the *Central Limit Theorem*. Formally, the theorem can be stated as follows.

Theorem 3B: Central Limit Theorem

Let $\bar{X}_n$ be the mean of a random sample of size n from a distribution $f(x)$ with the mean μ and variance σ^2. Then the distribution of the random variable Z defined by

$$Z = \frac{\bar{X}_n - \mu}{\sigma/\sqrt{n}} \tag{3.15}$$

approaches $N(0, 1)$ as n increases to infinity.

The remarkable thing about this theorem is the fact that it holds true whatever the distribution of X, provided it has a finite variance. The importance of the theorem, from the viewpoint of applications, is that the mean $\bar{X}$ of a random sample from any distribution is approximately normal with mean μ and variance σ^2/n if the sample size is large. This is equivalent to saying that $\sum_{i=1}^{n} X_i$ is approximately normal with mean $n\mu$ and variance $n\sigma^2$. [See (2.20) and (2.25).] In other words, the variable defined by

$$Z = \frac{\sum_{i=1}^{n} X_i - n\mu}{\sqrt{n}\,\sigma}$$

has an approximate standard normal distribution for large n.

The Central Limit Theorem can be proved mathematically although the proof is beyond the scope of this book. Even so, the tendency toward the standard normal distribution by Z, defined in (3.15), can be visually illustrated using an example. Let $\bar{X}_n$ be the sample mean of the points in n independent dice. The mean and variance of $\bar{X}_n$ are given by $\mu = 3.5$ and $\sigma^2 = (\frac{35}{12})/n = 2.917/n$. The exact distribution of $Z = (\bar{X}_n - 3.5)/(\sigma/\sqrt{n})$ for $n = 1, 2, 3, 4$ is plotted in Fig. 3.8. The rather rapid progress toward normality is striking and illuminating.

Even though the method of determining the distribution of Fig. 3.8 can not be described here, some readers may find it is a very interesting exercise to demonstrate a similar result using a computer simulation. Of course, the approach toward normality may be relatively slow with some distributions while rapid with others.

The practical aspect of the Central Limit Theorem is that it enables us to analyze and solve many problems which would be either impossible or difficult by other methods. We have already seen applications of this theorem where binomial as well as Poisson variables are approximated by a normal variable. In many applications we are dealing with analysis involving the mean or the sum of many independent observations. To mention an example,

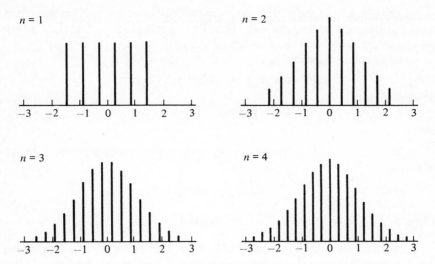

Fig. 3.8   Standardized distribution of sample mean of points in n dice

consider the electricity consumption in a city per day. It is the sum of the use by a large number of customers. This situation applies also to water consumption, gasoline consumption, total sales of a store, and so on. We shall consider a few applications of the Central Limit Theorem below. Many more subtle applications of the theorem will be found later in the text.

Example 3.8 The quantity of daily demand in pounds for a certain stock item is given by an unknown distribution with mean and standard deviation of 200 lb and 10 lb, respectively. Calculate the probability that the monthly demand is greater than 5100 lb assuming 25 business days in a month.

Let X_i be the demand on the ith day where $i = 1, 2, \ldots, 25$. From the Central Limit Theorem, the sum $\sum_{i=1}^{25} X_i$ can be assumed to have a normal distribution with the mean equal to $n\mu = (25)(200) = 5000$ and the variance $n\sigma^2 = (25)(100) = 2500$. It follows that

$$P(\sum X_i > 5100) = P\left(\frac{\sum X_i - 5000}{50} > \frac{5100 - 5000}{50}\right)$$

$$= P(Z > 2) = 0.023.$$

Thus, the chance of the monthly demand exceeding 5100 lb is about 2.3 %.   ▲

Note that the above problem is not tractable by any exact method because of the unknown distribution associated with the quantity of daily demand.

Example 3.9 Suppose that an unknown distribution with an unknown mean, μ, has a variance equal to 1. How large a sample must be taken before we can be 99% certain that $\bar{X}$ is within 0.5 of μ?

The problem requires that

$$P(-0.5 < \bar{X} - \mu < 0.5) \geq 0.99;$$

or equivalently,

$$P\left(\frac{-0.5}{1/\sqrt{n}} < \frac{\bar{X} - \mu}{1/\sqrt{n}} < \frac{0.5}{1/\sqrt{n}}\right) \geq 0.99.$$

Using the Central Limit Theorem, assuming that the sample size is going to be large,

$$\frac{0.5}{1/\sqrt{n}} \geq 2.576.$$

Thus, $n \geq 26.5$, or we need 27 observations. ▲

Again, no exact method could have been used to solve the above problem. The next example illustrates a problem that can be analyzed exactly (by more advanced techniques not discussed in this text), but can be solved here using the Central Limit Theorem. The approximate solution will be more than adequate in practice.

Example 3.10 Suppose that 30 electronic devices, say $D_1, \ldots, D_{30}$, are used in the following manner: as soon as D_1 fails, D_2 is put into operation, and when D_2 fails, D_3 becomes operative, etc. It is known that the lifetime of D_i has an exponential distribution with parameter 0.004/hr. (a) What is the probability that the total time of operations of the 30 devices exceeds a year? (b) How many devices should be stored so that the probability is about 0.99?

Let T denote the total time of operation of the 30 devices. The mean and variance of D_i are 250 hr and 62,500 hr², respectively. [See Eq. (3.8).] According to the Central Limit Theorem, T can be assumed to have an approximate normal distribution with mean (30)(250) hr and variance (30)(62,500) hr². Thus, noting that one year consists of 8760 hrs, we have

$$P(T > 8760) = P\left(Z > \frac{8760 - 7500}{1369.31}\right) = P(Z > 0.92)$$

$$= 0.18.$$

To answer the second question, let n denote the required number of devices. Then n has to satisfy the equation

$$P\left(Z > \frac{8760 - 250n}{250\sqrt{n}}\right) = P(Z > -2.326) = 0.99.$$

Solving for n yields

$$250n - 581.5\sqrt{n} = 8760,$$

from which we have $n = 51.77$, or we need about 52 devices to last one year with the probability 0.99. ▲

Note that it is assumed in the Central Limit Theorem that random variables are independent. Occasionally, we encounter the situation when X_1, $X_2, \ldots, X_n$ are not independent. Although detailed discussion is beyond the scope of this book, it is known that $\bar{X}_n$ approaches a normal distribution if X's are nearly independent or if at least X_i and X_j, which are sufficiently apart, are nearly independent. See Ref. 6.

Incidentally, if $\bar{X}_n$ is the sample mean based on the sample size n from a normal distribution with the mean μ and the variance σ^2, then the variable

$$Z = \frac{\bar{X}_n - \mu}{\sigma/\sqrt{n}}$$

has the exact standard normal distribution regardless of the sample size n. This follows essentially from the fact that any linear combination of variables with normal distributions is also normally distributed.

Example 3.11 Consider the $\bar{X}_n$ based on the random sample of size n from the standard normal distribution. Find the probability that $\bar{X}_n > 0.44$ for $n = 4$, 9, and 16. $N(0,1) \qquad \mu = 0 \quad \sigma^2 = 1 \quad \therefore \sigma = 1$

For $n = 4$, we have

$$P(\bar{X}_4 > 0.44) = P\left(Z > \frac{0.44 - 0}{1/2}\right) = P(Z > 0.88) = 0.19.$$

Similarly, the probabilities for $n = 9$ and 16 are calculated to be 0.093 and 0.039, respectively. When $n = 1$, the probability is 0.33. ▲

3.7 Lognormal Distribution

This distribution is closely related to the normal distribution. The nonnegative continuous variable X is said to have a *lognormal distribution* if $\log X$ is normally distributed; that is, if X is of the form e^Y where Y is normally distributed. Thus, there exists a one-to-one correspondence between a lognormal and a normal distribution. The density function of a lognormal variable is given by

$$f(x) = \frac{1}{x\sqrt{2\pi}\sigma} e^{-1/2[(\log x - \mu)/\sigma]^2}, x > 0, \tag{3.16}$$

where μ and σ^2 are the mean and variance of $Y = \log X$.

The graphical sketch of the lognormal density function with different values of μ is shown in Fig. 3.9. The skewness of a lognormal distribution, for fixed μ, increases with increasing value of σ.

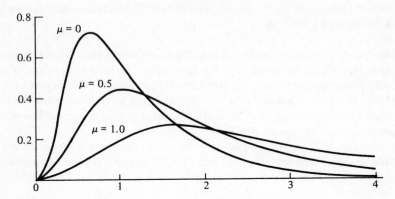

Fig. 3.9 Lognormal distribution with $\mu = 0$, $\mu = 0.5$, $\mu = 1.0$ when $\sigma = 1.0$

The mean μ_x and variance σ_x^2 of the lognormal variable X in terms of μ and σ^2 are

$$\mu_x = e^{\mu + \sigma^2/2}, \tag{3.17}$$

$$\sigma_x^2 = (e^{\sigma^2} - 1)e^{2\mu + \sigma^2}. \tag{3.18}$$

It is important to keep in mind the relationship given by the above two equations. Occasionally, the mean and variance, μ_x and σ_x^2 of the variable X with a lognormal distribution, are erroneously evaluated as $\exp(\mu)$ and $\exp(\sigma^2)$.

Because of the one-to-one relationship between a normal and a lognormal distribution, it is easy to evaluate the probability associated with a lognormal variable X. For example, we have

$$P(X < a) = P(\log X < \log a)$$

$$= P\left(Z < \frac{\log a - \mu}{\sigma}\right),$$

where μ and σ are the mean and variance of $\log X$, and Z has $N(0, 1)$.

The median of X is the number M_x such that $P(X \leq M_x) = 0.5$, or in terms of $\log X$, $P(\log X \leq \log M_x) = 0.5$. Since $P(\log X \leq \mu) = 0.5$, $\mu = \log M_x$ and, therefore, the median of X is

$$M_x = e^\mu. \tag{3.19}$$

We can see that the mean is larger than the median in the case of a lognormal distribution. From (3.17) and (3.19), we can obtain the following formulas which express the mean and variance of $\log X$ in terms of the mean and median of X:

$$\mu = \log_e M_x, \qquad \sigma^2 = 2 \log_e (\mu_x/M_x).$$

The lognormal distribution has been found to provide a satisfactory description of a wide variety of data in biomedical, socioeconomic, engineering, and other fields. It is used as the distribution of income, of weight of newborns, of cholesterol level, of survival time, and many others. As we will see in later chapters, many statistical methods require that the underlying distribution be normal, or Gaussian. Because of the relationship between a normal and a lognormal distribution, the analyses of data, whose underlying distributions are lognormal, can be performed after the logarithmic transformation of the data since the transformed observation has a normal distribution. (See Sec. 7.7.1.)

Example 3.12 Suppose that the life length of an electronic device is lognormally distributed with the median life length of 170 hr. It is known that the probability of the device operating at least 150 hr is 0.90. What is the standard deviation of the life length?

Let T be the life length of the device. The mean μ of $\log T$ is given by $\mu = \log 170 = 5.136$. It is known that

$$P(T > 150) = 0.90.$$

From this it follows that

$$P(\log T > \log 150) = P\left(Z > \frac{\log 150 - 5.136}{\sigma}\right) = 0.90.$$

Setting $(\log 150 - 5.136)/\sigma = -1.282$ and solving for σ, we obtain $\sigma = 0.161$. Substituting $\mu = 5.136$ and $\sigma = 161$ into (3.18), we get the desired result: $\sigma_x = 27.91$ hours. ▲

3.8 Other Distributions Related to the Normal Distribution

All the distributions described so far are used, among other applications, for modelling distributions of actual data in a wide variety of fields. The following three distributions, closely related to normal distributions, are rarely used as models. Nevertheless, they have important roles in other areas of statistical analyses and occur frequently. The applications of these distributions will appear in Chap. 6, 7, 8, and 9. In this section, we shall be concerned mainly with the definition of the three distributions.

3.8.1 t-Distribution

The *t-distribution* (or more precisely, Student's *t*-distribution) with parameter $v = 1, 2, \ldots$ is specified by the density function

$$f(x) = \frac{1}{\sqrt{v\pi}} \frac{\Gamma[(v+1)/2]}{\Gamma(v/2)} \left(1 + \frac{x^2}{v}\right)^{-(v+1)/2}, \quad -\infty < x < \infty \quad (3.20)$$

where $\Gamma(x)$ denotes the gamma function of x. (See Ref. 7 for the definition of the gamma function.) The parameter v in (3.20) is known as degree of freedom. The mean and variance of the distribution are

$$\mu = 0, \qquad \sigma^2 = \frac{v}{v-2} \quad \text{for } v > 2. \tag{3.21}$$

The t-distribution is symmetrical about $x = 0$ and resembles the standard normal density function especially when v is large. A sketch of the density function for several values of the degrees of freedom v is illustrated in Fig. 3.10.

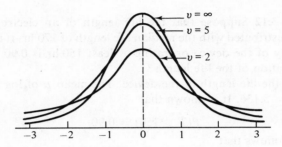

Fig. 3.10 t-distribution for $v = 2, 5$, and ∞

The relation between the t-distribution and the normal distribution will be discussed in Chap. 7 where the t-distribution is extensively applied.

The percentage points of the t-distribution t_α depend upon the number of degrees of freedom, and are given in Appendix Table A2. For example, the upper 5% points of the t-distribution with 2, 5, and 40 degrees of freedom are 2.92, 2.02, and 1.68, respectively. Note that the bottom row of Table A2 with infinite degrees of freedom gives entries identical to those percentage points of the standard normal distribution found in Appendix Table A1.

3.8.2 χ^2-Distribution

The *chi-square* (χ^2) *distribution* with parameter $v = 1, 2, \ldots$ is specified by the density function

$$f(x) = \begin{cases} \dfrac{1}{2^{v/2}\Gamma(v/2)} x^{(v/2)-1} e^{-x/2}, & x \geq 0 \\ 0, & x < 0 \end{cases}$$

where $\Gamma(x)$ denotes the gamma function of x.

The parameter v is called the degree of freedom of the distribution. The mean and variance of a χ^2-distribution are:

$$\mu = v, \qquad \sigma^2 = 2v.$$

The graphical sketch of the density function for several values of the degrees of freedom v is shown in Fig. 3.11.

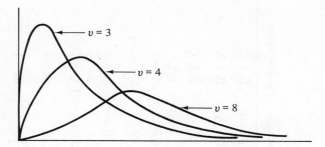

Fig. 3.11   Chi-square distribution for $v = 3$, $v = 4$, and $v = 8$

The χ^2-distribution is related to the normal distribution as follows: if $X_1, X_2, \ldots, X_v$ are v independent normal variables, each with zero mean and unit variance, the variable

$$\chi^2 = X_1^2 + X_2^2 + \cdots + X_v^2$$

has a χ^2-distribution with v degrees of freedom. The more practical version of the relation will appear in Chap. 7, where the χ^2-distribution is extensively applied.

The percentage points of the chi-square distribution are dependent on degrees of freedom and are given in Appendix Table A3. For example, the upper 5% points of the distribution with 3, 4, and 8 degrees of freedom are 7.81, 9.49, and 15.51, respectively.

3.8.3 F-Distribution

The *F-distribution with parameters* (m, n); $m = 1, 2, \ldots$ and $n = 1, 2, \ldots$ is specified by the density function

$$f(x) = \begin{cases} \dfrac{\Gamma[(m+n)/2]}{\Gamma(m/2)\Gamma(n/2)} (m/n)^{m/2} \dfrac{x^{(m/2)-1}}{[1+(m/n)x]^{(m+n)/2}}, & x \geq 0 \\ 0, & x < 0. \end{cases}$$

The parameters (m, n) are called the *degrees of freedom* of the F-distribution. The mean and variance are

$$\mu = \frac{m}{n-2} \quad \text{for} \quad n > 2$$

$$\sigma^2 = \frac{2n(m+n-2)}{m(n-2)^2(n-4)} \quad \text{for} \quad n > 4$$

The F-distribution involves two numbers of degrees of freedom, m and n, frequently referred to as the numerator and the denominator degrees of freedom, respectively. It is important to keep their order straight. Appendix Table A4 gives the percentage point F_α such that $P(F > F_\alpha) = \alpha$ for m numerator and n denominator degrees of freedom. For example, if $m = 5$ and

TABLE 3.3 Table of common distributions

Distribution of variable X	Form	Parameters	Mean	Variance	Possible values
Bernoulli	$p(x) = p^x(1-p)^{1-x}$	p	p	$p(1-p)$	$0,1$
Binomial	$p(x) = \dfrac{n!}{x!(n-x)!}p^x(1-p)^{n-x}$	n, p	np	$np(1-p)$	$0,1,2,\ldots,n$
Poisson	$p(x) = \dfrac{\lambda^x e^{-\lambda}}{x!}$	λ	λ	λ	$0,1,2,\ldots$
Geometric	$p(x) = p(1-p)^{x-1}$	p	$1/p$	$(1-p)/p^2$	$1,2,\ldots$
Exponential	$f(x) = \lambda e^{-\lambda x}$	λ	$1/\lambda$	$1/\lambda^2$	$(0,\infty)$
Normal	$f(x) = \dfrac{1}{\sqrt{2\pi}\sigma}e^{-\frac{1}{2}\frac{(x-\mu)^2}{\sigma^2}}$	μ, σ	μ	σ^2	$(-\infty,\infty)$
Lognormal	$f(x) = \dfrac{1}{x\sqrt{2\pi}\sigma}e^{-\frac{1}{2}\frac{(\log x-\mu)^2}{\sigma^2}}$	μ, σ	$e^{\mu+\sigma^2/2}$	$(e^{\sigma^2}-1)e^{2\mu+\sigma^2}$	$(0,\infty)$
Student's t	$f(x) = \dfrac{1}{\sqrt{\pi v}}\dfrac{\Gamma[(v+1)/2]}{\Gamma(v/2)}(1+x^2/v)^{-(v+1)/2}$	v	0 (for $v > 2$)	$v/(v-2)$(for $v-2$)	$(-\infty,\infty)$
Chi-square	$f(x) = \dfrac{1}{2^{v/2}\Gamma(v/2)}x^{(v-2)/2}e^{-x/2}$	v	v	$2v$	$(0,\infty)$
F	$f(x) = \dfrac{\Gamma[(m+n)/2]}{\Gamma(m/2)\Gamma(n/2)}(m/n)^{m/2}\dfrac{x^{(m/2)-1}}{[1+(m/n)x]^{(m+n)/2}}$	m, n	$\dfrac{m}{n-2}$	$\dfrac{2n(m+n-2)}{m(n-2)^2(n-4)}$	$(0,\infty)$

$n = 9$, we see that $F_{0.05} = 3.482$, whereas if $m = 9$ and $n = 5$, then $F_{0.05} = 4.773$. Readers may also verify that F_α with $(1, v)$ degrees of freedom is equal to t_α^2 where t_α is the 100α percentage point of the t-distribution with v degrees of freedom. Fig. 3.12 displays the F-distribution for several sets of degrees of freedom.

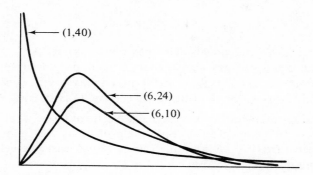

Fig. 3.12 F-distribution with $(m, n) = (6,24)$, $(m, n) = (6, 10)$, and $(m, n) = (1, 40)$ degrees of freedom

TABLE 3.4 Examples of random variables with the distribution given in Table 3.3

Bernoulli	The simple event with only two possible outcomes, called success and failure.
Binomial	The number of successes in n independent Bernoulli trials in which the probability of success at each trial is p.
Poisson	The number of events occurring in a unit time when the event occurs randomly at a mean rate λ per unit time; also the number of successes in n independent Bernoulli trials when n is large and p is small.
Geometric	The number of independent Bernoulli trials to obtain the first success when the probability of success is p.
Exponential	The waiting time required to observe the first occurrence of an event when the event is occurring randomly at a mean rate λ per unit time.
Normal	The sample mean based on a large number of observations.
Lognormal	e^Y where Y has a normal distribution.
Student's t	$(\bar{X} - \mu)/(s/\sqrt{n})$ where $\bar{X}$ is the sample mean of a normal distribution with the mean μ and s is the sample standard deviation.
Chi-square	$(n - 1)s^2/\sigma^2$ where s^2 and σ^2 are the sample and population variance of a normal distribution.
F	$(s_1^2/\sigma_1^2)/(s_2^2/\sigma_2^2)$ where s_1^2 and s_2^2 are the two independent sample variances of normal distributions with the variances σ_1^2 and σ_2^2, respectively.

3.9 Summary

The distributions described in this chapter are summarized in Table 3.3, and some examples of variables which have these distributions are given in Table 3.4.

PROBLEMS

1. Consider three individuals who contract a certain type of cancer; medical experience has shown that 30% of people contracting this disease do not survive. What is the probability that: (a) none of the three will survive, (b) all of them will survive, (c) at least one will live?

2. A certain manufacturing process yields electrical fuses, of which, in the long run, 10% are defective. Find the probability that in a sample of ten fuses, there will be (a) no defectives, (b) at least one defective, (c) no more than one defective.

3. An accident-insurance company finds that 0.1% of the population incurs a certain kind of accident each year. Assuming that the company has insured 10,000 persons selected randomly from the population, what is the probability that not more than three of the company's policyholders will incur this accident in a given year?

4. The incidence of measles in vaccinated children is approximated to be 25 per 100,000. (a) In a town of 1000 children, what is the probability of having two or less cases? (b) In a city of 10,000 children, what is the probability of having two or less cases? State your assumptions.

5. A radioactive source is observed during four time intervals of 6 sec each. The number of particles emitted during each period are counted. If the number of particles emitted can be described by a Poisson probability at a rate of 0.5 particles per sec, find the probability that (a) in each of the four time intervals three or more particles will be emitted, (b) in at least one of the four time intervals three or more particles will be emitted.

6. The average rate of murders in the city of St. Louis has been just about 0.7 per day. There had been three murders in one day. Is it unusual to have this many murders committed on just one given day? Discuss.

7. At a certain university the probability that a student, selected at random on a given day, will require a hospital bed is 1/5000. If there are 10,000 students, how many beds should the hospital have so that the probability that a student will lack for a bed is less than 2% (in other words, find K so that $P[X > K] \leq 0.02$, where X is the number of students requiring beds).

8. Let X be the number of particles emitted in t hours, and suppose that X has a Poisson distribution with parameter $\lambda = 30$. What is the probability that the time between successive emissions will be greater than 5 min?

9. The mean systolic blood pressure of American white males 18–44 years of age is 125 with a standard deviation of 13.7. If the blood pressure is normally distributed, what is the probability that a man of this age group, selected at random, has a blood pressure (a) above 147.5?, (b) below 110?, (c) between 110 and 140?

10. During the summer months, the average monthly electric bill in the area served by a certain electric company is $55 with the standard deviation of $15. (a) Assuming that the underlying distribution is normal, find the proportion of bills between $40 and $70. (b) Without assuming a normal distribution, estimate the probability that the sample mean based on 36 items in a sample will be less than $50.

11. By far the most widely prescribed medicine in the U.S. is the tranquilizer with the brand name "Valium." The published mean lethal dose (denoted by LD_{50}) of the drug is 720 mg/kg in mice. (a) If the standard deviation is 42.3 mg/kg, what percentage of a group of mice will die from a dose less than 650 mg/kg if it is assumed that the lethal dose has a normal distribution? (b) If it has a lognormal distribution?

12. The diameters of ball bearings are normally distributed with mean 20 mm and standard deviation 0.4 mm. The bearings with diameters less than 19 mm or greater than 21 mm are discarded as defectives. If bearings are sampled one at a time, what is the expected number of bearings sampled when the first defective appears? means geometric distribution

13. Suppose that 60% of a population has been found to belong to blood type "O." (a) Write, but do not compute, the probability that exactly 100 persons in a group of 200 belong to this blood type. (b) Compute this probability by means of the normal approximation.

14. Suppose that X_i, $i = 1, 2, \ldots, 50$ are independent random observations each having a Poisson distribution with parameter $\lambda = 0.05$. Let $S = X_1 + X_2 + \cdots + X_{50}$. (a) Using the Central Limit Theorem, evaluate $P(S \geq 2)$. (b) Compare the above answer with the exact value using the following *reproductive property* for the Poisson distribution: if $X_1, X_2, \ldots, X_n$ are independent variables with Poisson distributions with parameters $\lambda_1, \lambda_2, \ldots, \lambda_n$, then $S = X_1 + X_2 + \cdots + X_n$ has a Poisson distribution with parameter $\lambda_1 + \lambda_2 + \cdots + \lambda_n$.

15. It is found that in a certain city in the South, the annual family income has an approximately lognormal distribution, with the mean $12,500

and the median $11,000. (a) Find the proportion of families with incomes less than $6000. (b) Find the proportion with incomes $30,000 or greater.

16. The weight loss by evaporation of a certain packaged product has a mean of 7 grams with a variance of 1 gram. If three packages are selected at random from a lot, what is the probability that at least one shows a loss of 6 grams or more? (Make an appropriate assumption about the distribution of the weight loss.)

17. The chemical, "Tergitol S-9," washes the oil from birds' feathers, and, without the protecting oil, birds die from the cold. The probability of a bird dying from the cold is 0.95 if the temperature is below 20° and only 0.5 if it is below 30° but above 20°. The chemical was sprayed on a blackbird population of about 100,000 in a Kentucky town.

    (a) If the temperature is just below 30° (but above 20°), determine the probability that between 50,000 to 91,000 birds will be killed.
    (b) Suppose the forecast is for an 80% chance of being below 20°, and a 20% chance of being below 30° but above 20°. Calculate the probability that less than 95,000 birds will be exterminated.

18. The number of oil tankers arriving at a certain refinery each day has a Poisson distribution with the average of two (per day). The refinery can service only three tankers a day, and the tankers in excess of three must be sent to another refinery. Determine the chance of having to send tankers away on a given day.

19. Suppose that x_1 and x_2 are the two independent observations made from a normal distribution $N(1, 4)$. Find the probability:

    (a) $P(2x_1 - 2 < 4 \text{ and } x_2 < 5)$,
    (b) $P[(x_1 + x_2)/2 > 1 + \sqrt{2}]$.

*20. Let $\bar{X}$ denote the sample mean based on n observations from a normal distribution $N(\mu, \sigma^2)$. Determine n so the probability that the $\bar{X}$ will differ from μ by more than σ is at most equal to 0.01.

21. The chance for a successful launching of a spacecraft is 80%, and launching is continued until one successful shot is made. Suppose that each launching costs one million dollars and, in addition, a failure results in an additional cost of 0.25 million dollars. (a) What is the probability that more than three shots are required to make a successful launching? (b) Evaluate the expected cost of the program.

22. A large survey conducted in the St. Louis metropolitan area showed that the weight of 17-year-old white girls had a mean of 125 lb and a standard deviation of 20 lb, whereas for 17-year-old white boys, the mean was 153 lb with a standard deviation of 25 lb. Assuming that the above numbers are values of true parameters, answer the following

questions: (a) If one girl and one boy are each randomly selected, find the probability that the girl is between 115 and 105 lb and (simultaneously) the boy is heavier than 178 lb. (b) What is the probability that the girl is over 135 lb if it is certain that she weighs at least 125 lb?

*23. Suppose that X has a geometric distribution. Show that for any two positive integers i and j,

$$P(X > i + j \mid X > i) = P(X > j).$$

This equality states that if a success has not occurred in the first i trials, then the probability that it will not occur in the next j attempts is the same as the probability that it will not occur in the first j trials.

4

Organization of Data
and Descriptive Statistics

4.1 Introduction

As defined in Chap. 2, a population is the totality of elements or subjects that have one or more characteristics in common. The set of data that is taken from a population for the purpose of obtaining information about the specified property or characteristic of the population is called a *sample*. In brief, statistical methods may be considered procedures for drawing a conclusion on the population about the characteristic or the property of interest with a probabilistic interpretation.

The problem of how to take a sample from a population, so that valid conclusions about its characteristics may be obtained, is important. The essential requirement for a sample is that it be a *random sample*. We say that it is a random sample if each member of the population has an independent and equal probability of being selected in a sample. One intuitive reason for taking a random sample is that, in general, a sample selected only in this manner correctly represents the population under consideration. Furthermore, all statistical methods are based on the assumption of random sampling. Therefore, whenever we speak of a sample hereafter, we shall assume that it is a random sample from some population, even though the word random is not used explicitly.

4.2 Remarks on Sample and Population

The following discussion may clarify the preceding comment about the importance of random sampling. Consider the situation depicted in Fig. 4.1, in which B represents a subpopulation of a larger population denoted by A. Suppose that a random sample is obtained from B. Although such a sample is also a sample of A, it is not a random sample of A; it is a random sample from B. Unless A is homogeneous, such a sample is not likely to represent A but only B. Consequently, any conclusion reached from statistical analysis based on the sample will apply to B only, a subset of A. As an illustration, a sample of individuals surveyed at a shopping center of a city may be a random sample of a certain population, but certainly not of the entire city. Some readers may recall that in 1948 Truman won the presidential election when numerous polls predicted Dewey's victory. The polls at that time were heavily biased with urban voters and were not based on the random sample representing the population of all the voters.

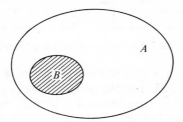

Fig. 4.1 Subpopulation B contained in population A

In many practical applications, it is not always easy to define a population under study or of interest. If the population is defined too narrowly, it may not be of sufficient generality or interest, while if the population is too broadly defined, the problem of random sampling would become more difficult and sometimes almost impractical. In this regard, we should recognize the many practical common situations when we don't have the luxury of taking a random sample from a well-defined population. Often it is impractical to take a random sample from a population. Sometimes we are confronted with an inverse situation when we are given a sample before a population is properly defined. It is important to construe the appropriate population that is represented by a sample, because the scope of any conclusion reached from analysis of the sample may be pertinent only to a specified population.

4.3 Processes in Statistical Inference

The part of statistical methods dealing with organizing and summarizing data is called *descriptive statistics.* The part concerned with drawing conclusions about a specified characteristic of the parent population is called *statistical inference.* Modern statistical method is mainly concerned with inference. Even so, in order to make an inference about the population, the descriptive part of statistics is usually required, and therefore, descriptive statistics may be considered as a preliminary part of statistical inference. The following simple diagram in Fig. 4.2 illustrates the usual processes in statistical inference.

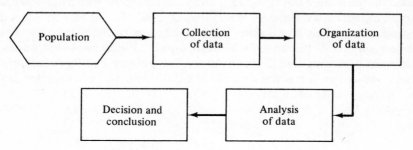

Fig. 4.2   Processes in statistical inference

Collection of data is a very important part of a statistical procedure, and we have briefly touched upon the subject at the beginning of this chapter. However, in the interest of conciseness and brevity, we shall not go into a detailed discussion of it, but shall consider the treatment of data already available, and in particular, descriptive statistics in this chapter.

When a considerable amount of data is available, a clear description is important for understanding, interpretation, and sometimes further analysis of such data. Two methods of describing data are in common use. One method involves a summarized presentation in tabular form organized from the original data. The other method consists of presenting the data in pictorial form—graphs, charts, or other similar representations. Before we proceed, it is important to remember that it is the description of the population characteristic that we wish to represent. The reason for this is that the data is obtained from a sample which is a subset of the population and the population characteristic is our main interest.

4.4 Sample Frequency Distributions

Statistical data as a rule consist of a set of measurements or counts. As an example of simple statistical data, consider Table 4.1, which presents the size of 40 families in a city. The observation, or strictly speaking, the

TABLE 4.1 Family size

5	2	4	2	8	2	4	3
4	7	2	4	2	4	7	5
3	3	6	3	1	5	5	4
7	5	5	4	3	5	4	6
5	4	4	3	6	6	4	4

variable X that we observe, can be either discrete or continuous. The family size considered here is an example of observations made on a discrete variable. In practice, discrete variables may be considered as variables whose possible values are integers; hence, they involve counting rather than measuring on a continuous scale.

Data such as that of Table 4.1 are usually summarized in the form of a *sample frequency distribution* as shown in Table 4.2. The relative frequencies in the third column are the frequencies divided by the sample size. A table consisting of the first and third columns is referred to as a *sample relative frequency distribution*, which gives the proportion of observations for each value of X. Intuitively, a sample relative frequency distribution is an empirical analogue of the true theoretical distribution of variable X.

TABLE 4.2 Frequency distribution of family size

Family size X	Frequency	Relative frequency
1	1	$\frac{1}{40}$ = .025 *freq*
2	5	.125 n_{Total}
3	8̶ 6	.̶1̶2̶5̶ .15
4	1̶3̶ 12	.̶3̶2̶5̶ .3
5	8	.200
6	4	.100
7	3	.075
8	1	.025

$\Sigma = 40$

In constructing the sample frequency distribution, we may wish to use some sort of grouping. For example, in Table 4.2, we could pool the family size into the class of 1 or 2, 3 or 4, 5 or 6, etc. Grouping is necessary for constructing such distributions for continuous data. As an illustration, consider the records of the weights of 45 newborns in a certain hospital, as presented in Table 4.3.

TABLE 4.3 Weight of newborns (lb)

6.5	5.8	7.1	6.9	7.8	6.9	7.8	8.6	7.0
9.5	6.9	7.6	7.7	6.8	7.3	7.2	6.5	8.1
6.4	8.2	6.8	6.6	8.7	7.5	7.2	7.2	6.3
7.9	6.2	7.2	7.1	7.6	9.8	9.2	6.7	6.9
6.4	7.5	8.3	7.3	6.7	6.4	8.4	6.4	7.7

In practice, continuous observations are recorded or rounded to a fixed number of decimal places, as in Table 4.3. The sample frequency distribution for the data is constructed in Table 4.4 by grouping the weight into class intervals of 5.51–6.00, 6.01–6.50, etc. Note that each of the 45 measurements in Table 4.3 lies inside one of the nine intervals. Each measurement which falls inside one of the intervals is conveniently approximated by the midpoint of that interval. In almost every instance, it is preferable that the class intervals be of equal length. The number of intervals is arbitrary but usually chosen to be between about 6 to 15, with the smaller number for smaller quantities of data.

TABLE 4.4 Frequency distribution of weight of newborns

Weight (lb)	Midpoint	Frequency	Relative frequency
5.51– 6.0	5.75	1	.044
6.01– 6.5	6.25	6	.178
6.51– 7.0	6.75	11	.244
7.01– 7.5	7.25	10	.222
7.51– 8.0	7.75	7	.089
8.01– 8.5	8.25	4	.089
8.51– 9.0	8.75	2	.044
9.01– 9.5	9.25	3	.067
9.61–10.0	9.75	1	.022

4.5 Graphical Representation of Sample Frequency Distribution

A frequency distribution is easier to visualize if it is displayed graphically. For discrete data, the representation is usually made by means of a *bar diagram*. The bar diagram for Table 4.2 is shown in Fig. 4.3. For continuous data, a more common form of graph is a *histogram*. The histogram for Table

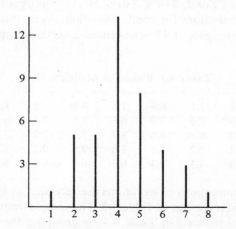

Fig. 4.3   Bar diagram for Table 4.2

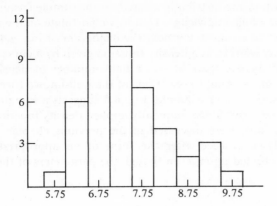

Fig. 4.4   Histogram for Table 4.4

4.4 is shown in Fig. 4.4, in which each interval is represented by its mid-point. Either the actual frequencies or the relative frequencies, whichever is desired, can be shown in graph form.

4.6 Relation Between Sample Frequency Distribution and Theoretical Distribution

It is important to realize the relation between a sample frequency distribution and the corresponding population, or theoretical, distribution. A theoretical distibution can be conceived as a limiting form of the sample relative frequency distribution as the sample size increases to infinity. In other words, a sample frequency distribution is an empirical estimate of the theoretical population distribution corresponding to it. For this reason, a sample distribution is sometimes called an empirical distribution.

In most statistical problems, the sample size may not be large enough to determine the theoretical distribution with much precision. However, on the basis of experience, theoretical considerations, or information in the sample, we can usually conceive and assume the functional form of the theoretical distribution involved. For example, consider the distribution of the variable X representing the sum of the number of points showing on two dice. In this case, theoretical reasoning leads us to determine the theoretical population distribution, without sampling, as given in Sec. 2.2. Experience with sample distributions of various linear measurements, such as the height of corn, diameters of steel pipes, etc., shows they tend to possess the normal distribution, at least approximately. We also know that the distributions of a sample mean tend to have a normal distribution due to the Central Limit Theorem. Now, consider the data presented in Table 4.3. In this case, the variable X, which determines the population chracteristic, is unknown, and we have to idealize the form of the distribution based on the sample frequency distribution or other available knowledge. That is, we postulate on the model for the distribution. As an example, consider the histogram of Fig. 4.4 again. From its shape we may want to fit a density function given by a curve as illustrated in Fig. 4.5. Of course, there are an infinite number of density functions. Usually, however, we want to select one of the well-known forms in order to facilitate the analysis. For example, Fig. 4.5 suggests a lognormal distribution as a suitable form. Some frequently applied density functions, as well as frequency functions, were discussed in the previous chapter. Most of the statistical analyses proceed after the form of an appropriate theoretical distribution is decided upon, even though the parameters of the distribution are unknown.

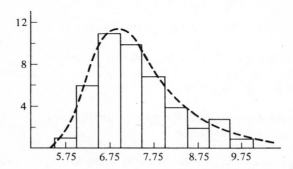

Fig. 4.5 Histogram for the weight of newborns and a lognormal distribution

4.7 Measure of Location

To understand and visualize the population distribution on the basis of a sample, we construct a frequency distribution and histogram. It is useful to summarize the data further by giving certain measures which empirically describe important features of the distribution. The first is a measure of location or center of the distribution, and the second is that of dispersion or variation.

The most commonly used measures of the center of the distribution, calculated from the sample, are the mean and median. The *sample mean* is what is frequently called the *average*. Given $X_1, X_2, \ldots, X_n$, the values of n observations, the mean denoted by $\bar{X}$ is

$$\bar{X} = \frac{\sum_{i=1}^{n} X_i}{n}. \tag{4.1}$$

For example, if three men weigh 130, 180, and 150 lb, the mean weight is $\bar{X} = 153.3$ lb.

For grouped data like Table 4.4, it is necessary to interpolate to calculate the mean. Let y_i be the mid-point of the ith group, and let f_i be its frequency. Then, it is easy to see that the mean, $\bar{X}$, is given by

$$\bar{X} = \frac{\sum f_i y_i}{n} \tag{4.2}$$

where $n = \sum f_i$. (See Prob. 6.)

We should note that the sample mean estimates the true mean, μ, of the distribution of a population under investigation. (See Theorem 2A.)

Example 4.1 Consider the frequency distribution of Table 4.4. The sample mean, using (4.2), is

$$\bar{X} = [(1)(5.75) + (6)(6.25) + \cdots + (1)(9.75)]/45 = 7.38.$$

Therefore, the mean weight based on the sample of 45 newborn babies is approximately 7.38 lb. Incidentally, the sample mean calculated directly from the data of Table 4.3 is 7.35 lb. ▲

The measure of location used next most frequently to the mean is the *sample median* which can be denoted by m_x. The median of a set of n observed values is the middle number in the array. Roughly, it is the value given so that there are as many observations whose values are above the value as there are observations whose values are below the value. If n is an odd number, the median is given by the $(n + 1)/2$ number in the array. If n is an even number, the median is defined as the average of the numbers in the array with ranks $n/2$ and $(n + 2)/2$. For example, if three men weigh 130, 180, and 150 lb, the median is 150 lb. If a fourth man weighs 190 lb, then the median computed from the four numbers is 165 lb. It should be noted that, if n is large, calculation of the median is not an easy task because of the labor involved in ranking the data.

One measure less frequently applied than the mean and median is the *geometric mean*, which can be used only when all observations are positive and nonzero. For a set of n positive numbers $X_1, X_2, \ldots, X_n$, the geometric mean, denoted by g_x, is defined as

$$g_x = (X_1 \cdot X_2 \cdots X_n)^{1/n}. \tag{4.3}$$

Since $\log(g_x) = (\sum_{i=1}^{n} \log X_i)/n$, the geometric mean can be conveniently calculated by

$$g_x = e^{(\sum \log X_i)/n}. \tag{4.4}$$

The geometric mean is recommended for data for which the ratio of any two consecutive numbers is constant or nearly constant. This occurs, for example, in data representing the amount of investment which is increasing at compound interest.

Example 4.2 An amount of $1000 is invested at 6% interest on January 1 of this year. If interest is added annually on January 1, determine the average amount of investment between January 1 of this year to December 31 of the fourth year, using the geometric mean.

The amount of investment on January 1 can be computed as:

Year	Amount (X)	$\ln X$ / $\log X$
1	1000.00	6.908
2	1060.00	6.966
3	1123.60	7.024
4	1191.02	7.083
		27.981

The geometric mean is $g_x = 1091.44$. Incidentally, we have $\bar{X} = 1093.66$, and the median is $m_x = 1091.80$ for the above data. ▲

The most important single measure of location calculated from a sample is the mean. It is easy to calculate, easy to define, takes all measurements into consideration, and is, in general, very reliable. However, the mean is sensitive to extreme values. In general, if the distribution is very much skewed, the mean may not provide the best description. Under these circumstances, the median can be more descriptive of the central tendency of the distribution because the median is not influenced by extreme measurements. As an example, the per capita income, which is the mean income per person for the U.S. in 1976, was about $7060, while the corresponding figure for Kuwait was $11,510. Obviously, the median would better describe the living standard, although the per capita income is used more often in practice for this purpose. The situation may be very much the same when we want to describe the average number of white blood cells per cu mm in a sample of patients or the mean survival time of a certain component. Notwithstanding the advantage of the median over the mean in the example given, the median is very difficult to calculate even with the help of a computer if n is large, as has been mentioned already. Sometimes, for this reason, the geometric mean is used as a substitute for the median in a positively skewed distribution (which has a long tail to the right), because the geometric mean is much easier to calculate, especially by a computer, than the median if the sample size is large. For many practical situations, the geometric mean and the median may be close. It can be shown that the geometric mean is never greater than the mean.

4.8 Measure of Variation

In the last section, we noted that one way to describe a set of observed values is by giving a measure of location. However, the mean or median does not describe the amount of dispersion or variation among the observed

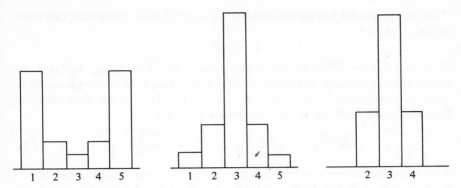

Fig. 4.6 Three frequency distributions with the same mean but different variations

values. This fact is illustrated in Fig. 4.6 where three sample frequency distributions with very different dispersions have identical means.

The problem of measuring variation arises frequently in various fields of science, and the problem will be investigated more fully in a later chapter. In this section, we shall simply introduce and describe some measures of the dispersion or variation of the sample. The first measure of variation is the _range_, which is defined as the difference between the largest and smallest values in the set. It is a well-known and popular measure due to its simplicity, and is found to be useful in some fields of applications such as quality control work. The range has two disadvantages. First, as the sample size is increased, the range generally tends to become larger; it depends on the number of observations. Secondly, as might be suspected intuitively, the range is not a very efficient measure because it depends solely on two extreme values. For example, in Fig. 4.6, the first two frequency distributions have the same range of 4, but the second distribution exhibits less variation as its values are more concentrated about the mean. As a measure of dispersion, the range is usually recommended when the sample size is less than 10.

The second measure of variation is the *mean deviation,* which is defined as

$$\text{Mean deviation} = \frac{\sum_{i=1}^{n} |X_i - \bar{X}|}{n}. \tag{4.5}$$

This measure of variation is of limited use mainly because analyses based on the mean deviation can be rather complicated.

The most important and most commonly used measure for the variation of observations is the *sample variance,* or the *sample standard deviation.* Suppose the data consist of n observed values, $X_1, X_2, \ldots, X_n$, whose mean is $\bar{X}$. The sample variance denoted by s^2 is given by

$$s^2 = \frac{\sum (X_i - \bar{X})^2}{n - 1}. \tag{4.6}$$

The sample standard deviation denoted by s, or SD, is the positive square root of the variance:

$$s = \sqrt{s^2}.$$

Note that the larger the variation in a sample about its mean, the larger is the standard deviation. When the standard deviation is to be computed by a calculator (although some models have a built-in key for it), the following formulas could facilitate the computations involved.

$$s = \sqrt{\frac{\sum X_i^2 - (\sum X_i)^2/n}{n-1}}, \quad \text{or} \quad s = \sqrt{\frac{\sum X_i^2 - n\bar{X}^2}{n-1}}. \tag{4.7}$$

We should remember that the variance or the standard deviation can never be a negative number. In this regard, we should avoid excessive round-off errors when computing these quantities.

As a simple descriptive measure of the variation based on a sample, the standard deviation s is often preferred over the variance s^2, since the unit of s is the same as that of the original observations, while s^2 is expressed in terms of the square of the unit. On the other hand, s^2 is used more often in analyses regarding variation in the population. Note that s^2 and s estimate unknown σ^2 and σ, respectively, of a population distribution.

A quantity related to the sample standard deviation s is the *standard error of the mean* which is denoted by $s_{\bar{x}}$, or SE, and given by

$$s_{\bar{x}} = s/\sqrt{n}. \tag{4.8}$$

It is simply the standard deviation of the sample mean, and is commonly used as a measure of precision or reliability of the sample mean. (See Theorem 2A.) Thus, it is important to remember s provides a measure of the variation of individual values, while $s_{\bar{x}}$ is an estimate of the variation of a sample mean.

4.9 Coefficient of Variation

When it is desired to compare the variability of two or more distributions based on the samples, neither the variance nor the standard deviation may be used unless the sets of data have the same unit. If the sets of observations have different units, the *cofficient of variation V* would be useful for this purpose. It is calculated as

$$V = \frac{s}{|\bar{X}|}, \tag{4.9}$$

noting that it is undefined if $\bar{X} = 0$. The coefficient of variation is independent of the unit of measurement, and can be used to compare the relative variations of two or more sets of data regardless of the unit used in the observations. The underlying rationale is that the value of V is not altered if each observation in a sample is multiplied by a constant number, and that it is a unitless number.

4.10 Summary

In this chapter, we have described the method constructing the sample frequency distribution and two broad measures of the sample distribution: namely, the center and the variation. The most important and most frequently used measures are the mean and variance or the standard deviation which are calculated by the following formulas. In addition to the formulas, it is important to remember what they estimate in a population distribution.

	Sample statistic	Formula	Population parameter estimated
Mean	$\bar{X}$	$\sum X_i/n$	μ
Standard deviation	s	$\sqrt{\dfrac{\sum (X_i - \bar{X})^2}{n-1}}$	σ

It is a common practice for investigators to report the sample mean and standard deviation with the format $\bar{X} \pm s$ together with the sample size in summarizing data. In this regard, we frequently encounter results expressed as $\bar{X} \pm s_{\bar{x}}$, so it is important to indicate which one is meant. If we wish to express the sample mean along with its reliability, the format $\bar{X} \pm s_{\bar{x}}$ is to be preferred, whereas $\bar{X}$ and s should be presented if the objective is to describe the location and spread of the sample distribution. The latter may be the more common situation. For example, Table 4.5 summarizes the data on the blow observed from three baby grey whales. (From 1975 *Science* 190, p. 908.)

TABLE 4.5 Mean and SD of the blow. Abbreviations: *E*, expiration; *I*, inspiration; and *n*, the number of observations

Body length (m)	Total volume (liters) $\bar{X} \pm s$	n	Duration (sec) $\bar{X} \pm s$	n
4.77				
E	24 ± 7.5	26	0.54 ± 0.13	25
I	26 ± 7.0	31	0.40 ± 0.11	30
5.78				
E	38 ± 14.1	17	0.49 ± 0.19	18
I	29 ± 11.5	14	0.38 ± 0.10	12
5.21				
E	32 ± 12.2	6	0.41 ± 0.09	5
I	18 ± 6.7	3	0.38 ± 0.12	4

PROBLEMS

1. Discuss a possible method of selecting a random sample of size n from a population of eligible voters in your city.

2. Suppose that a random sample of size n_1 is obtained from a certain population. Now suppose the sample size n_1 is too large, and it is decided to select randomly a smaller sample of size n_2 from n_1. Is the final sample a random sample of the initial population?

3. In the closing month of a 1976 presidential primary election, polls taken a week apart by two well-known pollsters indicated a marked difference in the estimated proportion of the electorate intending to vote for a certain candidate. Cite three possible factors which might have contributed to the seeming discrepancy.

4. The salaries of a sample of 80 engineers in a large company are shown below. (a) Construct a histogram for the data. (b) Suggest a suitable distribution from those studied in Chap. 3 to describe the distribution of salaries.

Salaries ($)	Frequency
13,001–15,000	10
15,001–17,000	25
17,001–19,000	18
19,001–21,000	12
21,001–23,000	9
23,001–25,000	6

5. The annual rainfall in inches at a certain city for a 90-year period is shown below. (a) Construct a bar diagram for the data. (b) Name a suitable distribution from those studied in Chap. 3.

Rainfall	Frequency	Rainfall	Frequency
7.51–10.5	12	22.51–25.5	14
10.51–13.5	10	25.51–28.5	3
13.51–16.5	15	28.51–31.5	0
16.51–19.5	19	31.51–34.5	4
19.51–22.5	12	34.51–37.5	1

6. When the data are arranged in a frequency distribution, let y_i be the mid-point of the ith interval, and let f_i be the frequency. Show that the sample mean and standard deviation are approximated by

$$\bar{X} = \frac{\sum f_i y_i}{\sum f_i}, \qquad (4.10)$$

$$s = \sqrt{\frac{\sum f_i y_i^2 - (\sum f_i y_i)^2/n}{n-1}}, \tag{4.11}$$

where $n = \sum f_i$.

7. Using formulas (4.10–4.11), calculate the mean and standard deviation for the data given in Prob. 4.

8. A random sample of 20 households was taken from a town consisting of 10,000 households. The number of persons per household in the sample were as follows:

$$4, 3, 4, 3, 3, 5, 2, 5, 6, 3, 3, 4, 3, 2, 7, 4, 3, 5, 4, 2.$$

(a) Estimate the total population of the town. (b) Calculate the median of the sample. Can we estimate the population using the median? If not, what can be said about the town from the calculated median? (c) Calculate the standard deviation of the above numbers, and explain what it estimates in the town.

9. Determine the coefficient of variation for the data of Prob. 5.

10. For the data given in Table 4.3, calculate the geometric mean.

*11. Show that the geometric mean is never greater than the mean. In other words, prove that $g_x \leq \bar{X}$.

12. Show that the value of the coefficient of variation, V, does not change if each observation in a sample is multiplied by a constant number, either positive or negative. What would happen to V if a constant is added to each observation?

13. A sample of n_1 observations has mean $\bar{X}_1$ and standard deviation s_1, while another sample of size n_2 has mean and standard deviation $\bar{X}_2$ and s_2. Show that the mean and variance of the combined sample are

$$\bar{X} = \frac{n_1 \bar{X}_1 + n_2 \bar{X}_2}{n_1 + n_2},$$

$$s = \frac{(n_1 - 1)s_1^2 + (n_2 - 1)s_2^2 + n_1(\bar{X}_1 - \bar{X})^2 + n_2(\bar{X}_2 - \bar{X})^2}{n_1 + n_2 - 1}$$

$$\left[\text{Hint:} \ \sum_{i=1}^{n_1+n_2} (X_i - \bar{X})^2 = \sum_{i=1}^{n_1} (X_i - \bar{X}_1 + \bar{X}_1 - \bar{X})^2 \right.$$

$$\left. + \sum_{i=n_1+1}^{n_1+n_2} (X_i - \bar{X}_2 + \bar{X}_2 - \bar{X})^2 \right].$$

14. Show that the median m_x for grouped data can be approximated using the formula

$$m_x = b + c\frac{(n/2) - d}{f_m} \tag{4.12}$$

where
   b = lower boundary of the class that contains the median,
   c = length of the class that contains the median,
   d = number of observations whose values are less than the lower
        boundary of the group that contains the median, and
   f_m = frequency of the group that contains the median.

15. Using Eq. (4.12), calculate the medians for the data given in Table 4.4 and compare it with the median obtained from the raw data of Table 4.3.

16. Calculate the median for the grouped data of Prob. 4.

17. Show that the sample standard deviation for grouped data can be calculated by the formula

$$s = \sqrt{\frac{\sum (y_i - \bar{X})^2 f_i}{n - 1}} \qquad (4.13)$$

where y_i is the mid-point of the ith group with frequency f_i and $\bar{X}$ is the sample mean calculated by Eq. (4.2).

18. Using Eq. (4.13), calculate the sample standard deviation for the data given in Table 4.4 and compare it with the standard deviation calculated directly from Table 4.3.

*19. Show that $E(s^2) = \sigma^2$. [Hint: use the equality $\sum (X_i - \bar{X})^2 = \sum \{(X_i - \mu) - (\bar{X} - \mu)\}^2$ and Eq. (2.19).]

5

Statistical Inference:
Principles and Methods

5.1 Introduction

Statistical inference may be viewed as the science of making conclusions or decisions about the population based on the information obtained from samples. Two main categories of problems in inference are estimation and the test of hypothesis. The first is subdivided into point estimation and interval estimation. As an example, consider the breaking strength of a certain safety belt. Certainly, it will be of interest to determine the true mean strength of the belt. The *point estimation* deals with the problem of estimating the unknown parameter such as the true mean based on the sample. The number which estimates a parameter is called the *estimate*, and the function of the sample which can be used to calculate the estimate is called the *estimator*. For example, the sample mean denoted by $\bar{X}$ is an estimator, and 7500 lb is an estimate if this number happens to be the value of $\bar{X}$. It is very unlikely that the estimate will be precisely equal to the parameter. Thus, it may be a good idea to have an interval which is likely to contain the mean, with some specified assurance. We calculate the *confidence interval* from the sample for this purpose.

More often in practice, we are not concerned about the problem of estimating, but with deciding whether an unknown parameter is equal to a certain specified value. For example, we may wish to decide whether a proportion of defective items in a certain product equals a certain value such as 0.5%. The procedure which leads to accepting or rejecting specified statements about the population is called the *test of hypothesis*. The test of hypothesis dealing with a single population, as with the problem of the safety belt, is called the *one-sample test*. On the other hand, we may be dealing with two populations simultaneously. For example, we may wish to test whether the mean tensile strength of two different types of steel are equal or not. Such a test of hypothesis comparing two populations is called the *two-sample test*. The multi-sample test is analogously defined. In this chapter we shall study some principles and methods of estimation and tests of hypothesis.

*5.2 Basic Principles of Point Estimation

Suppose we have a random sample of n observations $X_1, X_2, \ldots, X_n$ from a population with distribution $f(x)$ which depends on the unknown parameter θ. We would like to estimate θ using some function of the sample. When we speak of estimating θ, we are speaking of estimating the unknown yet fixed value of θ. The estimator for θ is usually denoted by $\hat{\theta}$ and read θ hat. Naturally, there would be many alternative functions that could be considered for the purpose. For example, suppose $f(x)$ is a normal distribution and the mean μ is to be estimated. Then, among others, the sample mean, sample median, the midrange (the average of the largest and smallest numbers in a sample), etc., can be considered as candidates for $\hat{\mu}$, an estimator of μ.

Two critical questions in estimation problems are:

(a) What characteristics do we want a "good" estimator to possess?

(b) How do we decide whether one estimator is "better" than another?

First, it is important to note that an estimator is also a random variable since it is a function of a random sample, and so, like any other random variable, it has a distribution. Intuitively, a "good" estimator $\hat{\theta}$ is such that its distribution will be concentrated as closely as possible near the true value of θ. Let $\hat{\theta}_1$ and $\hat{\theta}_2$ be two possible estimators for θ with densities $g_1(\hat{\theta}_1)$ and $g_2(\hat{\theta}_2)$ as illustrated in Fig. 5.1.

In any given situation, the estimate using $\hat{\theta}_2$ may possibly be closer to θ than that using $\hat{\theta}_1$. On the average, however, the value based on $\hat{\theta}_1$ is likely

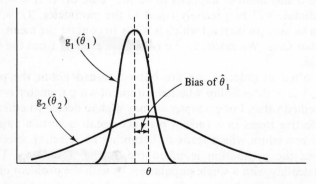

Fig. 5.1 Biased estimator $\hat{\theta}_1$ with small variance and unbiased estimator $\hat{\theta}_2$ with large variance

to be closer than that of $\hat{\theta}_2$. This statement can be made precise by saying that $\hat{\theta}_1$ has the smaller *mean square error* (MSE). The MSE of an estimator $\hat{\theta}$ is defined as

$$\text{MSE}\,(\hat{\theta}) = E(\hat{\theta} - \theta)^2.$$

This quantity can be shown to be

$$E(\hat{\theta} - \theta)^2 = E[(\hat{\theta} - E(\hat{\theta})) + (E(\hat{\theta}) - \theta)]^2$$
$$= \text{Var}\,(\hat{\theta}) + (E(\hat{\theta}) - \theta)^2,$$

since the cross product term $2E[(\hat{\theta} - E(\hat{\theta}))(E(\hat{\theta}) - \theta)] = 2(E(\hat{\theta}) - E(\hat{\theta}))$ $\cdot(E(\hat{\theta}) - \theta) = 0$. The term $E(\hat{\theta}) - \theta$ is called the *bias* of $\hat{\theta}$; it represents the expected magnitude of $\hat{\theta} - \theta$. Therefore, the MSE of $\hat{\theta}$ can be expressed as

$$\text{MSE}\,(\hat{\theta}) = \text{Var}\,(\hat{\theta}) + (\text{bias of } \hat{\theta})^2,$$

An estimator $\hat{\theta}$ is said to be *biased if $E(\hat{\theta}) \neq \theta$*, whereas it is called *unbiased if $E(\hat{\theta}) = \theta$*. In other words, an estimator is unbiased if and only if the mean of its distribution equals θ.

In Fig. 5.1 note that $\hat{\theta}_1$ is biased but $\hat{\theta}_2$ is unbiased. Even so, the MSE of $\hat{\theta}_1$ is smaller than that of $\hat{\theta}_2$, and so $\hat{\theta}_1$ may reasonably be regarded as a better estimator of θ than $\hat{\theta}_2$.

The comparison between two estimators $\hat{\theta}_1$ and $\hat{\theta}_2$ can be made quantitative by the *relative efficiency* of $\hat{\theta}_1$ to $\hat{\theta}_2$ which is denoted as RE $(\hat{\theta}_1/\hat{\theta}_2)$ and defined by

$$\text{RE}\,(\hat{\theta}_1/\hat{\theta}_2) = \text{MSE}\,(\hat{\theta}_2)/\text{MSE}\,(\hat{\theta}_1).$$

If the relative efficiency of $\hat{\theta}_1$ to $\hat{\theta}_2$ is less than one, then $\hat{\theta}_1$ is said to be less efficient than $\hat{\theta}_2$, whereas if it is greater than one, then $\hat{\theta}_1$ is the more efficient estimator.

Example 5.1 Consider a population whose density function has a normal distribution with the unknown mean μ and variance equal to 1. We wish to estimate the mean based on a sample of size 10. Consider four estimators: $\hat{\mu}_1 = \bar{X}$, $\hat{\mu}_2 = \bar{X} + 0.1$, $\hat{\mu}_3 = $ median, and $\hat{\mu}_4 = $ midrange. Table 5.1 presents the Var $(\hat{\mu})$, bias, and MSE of the four estimators. (The method of calculating the entries for $\hat{\mu}_3$ and $\hat{\mu}_4$ is beyond the scope of this text.)

Among the four estimators, $\hat{\mu}_1$ is clearly the best choice since it has the smallest MSE. If $\hat{\mu}_1$ is not available, then one may wish to choose $\hat{\mu}_2$, although

TABLE 5.1 Four estimators and their MSE when $n = 10$

Estimator	Var $(\hat{\mu})$	Bias	MSE
$\hat{\mu}_1 = \bar{X}$	0.100	0	0.100
$\hat{\mu}_2 = \bar{X} + 0.1$	0.100	0.1	0.110
$\hat{\mu}_3 = $ median	0.138	0	0.138
$\hat{\mu}_4 = $ midrange	0.186	0	0.186

it is biased and $\hat{\mu}_3$ and $\hat{\mu}_4$ are unbiased. We can calculate RE $(\hat{\mu}_3/\hat{\mu}_2) = 0.797$ and RE $(\hat{\mu}_4/\hat{\mu}_2) = 0.591$. Thus, $\hat{\mu}_2$ is more efficient than $\hat{\mu}_3$ or $\hat{\mu}_4$. ▲

Since the mean square error of an estimator can be used as the yardstick in assessing the quality of an estimator, it would seem a good idea to find estimators with the smallest possible MSE in estimating the parameters. It turns out, however, that such estimators with minimum MSE rarely exist in practice. A reasonable procedure is to restrict our attention to only unbiased estimators and then, among all possible unbiased estimators, select one with minimum MSE. Now note that if an estimator is unbiased, then its MSE coincides with its variance. Thus, finding an estimator with minimum MSE among unbiased estimators is equivalent to finding an estimator with minimum variance among unbiased estimators. Such an estimator is called the *minimum variance unbiased estimator* (MVUE). To be precise, $\hat{\theta}$ is called the MVUE of θ if it satisfies

(a) $E(\hat{\theta}) = \theta$,

(b) Var $(\hat{\theta})$ is less tnan the variance of any other unbiased estimator.

The concept of unbiased estimator and the MVUE can be illustrated by comparing the shooting performance of three different guns, A, B, and C. Accuracies of the three guns are depicted in Fig. 5.2. The analogy between selection of the estimators and choice of gun is self-explanatory. Suppose we are evaluating the shooting performance of the three guns. The target pattern of gun A is unbiased but its variance is very large, whereas gun B has small variance but is strongly biased. The hit pattern of gun C is unbiased and has small variance. Clearly, gun C is most desirable among the three guns.

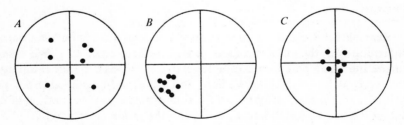

Fig. 5.2   Target pattern of three guns

Confining our attention only to unbiased estimators is not a bad idea. After all, unbiasedness is one of the desirable properties an estimator should have. In addition, unbiased estimators have the following desirable property: if we have two or more unbiased estimators, then their average is also unbiased. On the other hand, the average of biased estimators is again biased in general.

It can be shown that the sample mean $\bar{X}$ is the MVUE of the mean of normal, binomial, and Poisson distributions, and that the sample variance s^2

is the MVUE for the variance of a normal distribution. We shall not be able to study the method of finding the MVUE as such a study goes way beyond the level assumed in this text. Instead, we shall consider next an important method of obtaining an estimator which approaches the MVUE as the sample size increases.

*5.3 Method of Point Estimation

There are several general methods available that will yield estimators not necessarily the MVUE. The only method we shall discuss is the *maximum likelihood method* because it is both simple and the most useful.

To introduce the method, consider a very simple problem of estimating the parameter p of the Bernoulli frequency function:

$$f(x) = p^x q^{1-x}, \qquad x = 0, 1.$$

Suppose that an experiment yielded $X = 1$. Then it can be seen that $f(x)$ is maximized if p is estimated by $\hat{p} = 1$. In a sense, $\hat{p} = 1$ is the *most likely* value for p if $X = 1$. Similarly, if $X = 0$, then $f(x)$ is maximum if p is taken to be $\hat{p} = 0$. Now, consider the case when we have random samples $X_1, X_2, \ldots, X_n$. Then, the probability of observing such a sample is given by

$$\prod_{i=1}^{n} f(X_i) = f(X_1)f(X_2) \cdots f(X_n).$$

It gives the relative "likelihood" that X assumes the observed values $X_1, X_2, \ldots, X_n$, and it is referred to as the *likelihood function.* A reasonable idea is to estimate p by $\hat{p}$ which maximizes the likelihood function. This calls for estimating the parameter as if the parameter is the one giving the "best" explanation of the observed data where "best" means the one with the maximum likelihood.

Now, we formalize the maximum likelihood principle. Let $X_1, X_2, \ldots, X_n$ be a random sample from the random variable X which depends on a parameter θ. The likelihood function L of θ given $X_1, X_2, \ldots, X_n$ is defined as

$$L(X_1, X_2, \ldots, X_n; \theta) = \prod_{i=1}^{n} f(X_i) \tag{5.1}$$

where $f(X_i)$ denotes the density function of X if it is continuous and the frequency function if X is discrete, evaluated at X_i. The *maximum likelihood estimate* (MLE) of θ, say $\hat{\theta}$, based on a random sample $X_1, X_2, \ldots, X_n$ is that value of θ which maximizes $L(X_1, X_2, \ldots, X_n; \theta)$ which may be simply denoted by $L(\theta)$. In maximizing the likelihood function, the observed X's are considered as constants and θ as a variable.

Using calculus, the MLE is determined by taking the derivative of $L(\theta)$ with respect to θ and setting the derivative equal to zero. In practice, the computation is facilitated if we work with the logarithm (base e) of the likelihood $\ln L (X_1, \ldots, X_n; \theta)$ which is called the log-likelihood function. The maximizing value of θ for the likelihood function and the log-likelihood function is identical. Thus, the MLE is given by $\hat{\theta}$ which satisfies

$$\frac{\partial}{\partial \theta} \ln L(X_1, \ldots, X_n; \hat{\theta}) = 0.$$

Example 5.2 Suppose that a random sample $X_1, X_2, \ldots, X_n$ is drawn from the Bernoulli distribution

$$f(x) = p^x q^{1-x} \qquad x = 0, 1; 0 \leq p \leq 1.$$

The likelihood function is

$$L(p) = \prod_{i=1}^{n} p^{X_i} q^{1-X_i} = p^{\Sigma X_i} q^{n - \Sigma X_i},$$

and

$$\ln L(p) = \sum X_i \ln p + (n - \sum X_i) \ln (1 - p).$$

We set

$$\frac{\partial \ln L(p)}{\partial p} = \frac{\sum X_i}{p} - \frac{n - \sum X_i}{1 - p} = 0$$

and solving for p, the MLE is $\hat{p} = \sum X_i / n = \bar{X}$. ▲

An important property of the MLE is that it becomes the minimum variance unbiased estimator as the sample size increases to infinity. In other words, the MLE has the optimum property for large sample size. For a finite sample size, the MLE is not necessarily unbiased. Another important and convenient property of the MLE is that the MLE of a monotone function of θ is simply that function of the MLE. That is, if g is a monotone function, the MLE of $g(\theta)$ is given by $g(\hat{\theta})$ where $\hat{\theta}$ is the MLE for θ. Thus, for example, the MLE of $\log \theta$ and $1/\theta$ are $\log \hat{\theta}$ and $1/\hat{\theta}$, respectively.

Example 5.3 Suppose that the time T to failure of a component has an exponential distribution with parameter λ. A sample of size 10 gave these failure times: 1.6, 1.2, 0.8, 2.2, 1.4, 3.1, 1.5, 0.9, 0.6, and 0.7. Find the maximum likelihood estimate for the mean failure time.

The exponential density function is given by

$$f(t) = \lambda e^{-\lambda t} \qquad t \geq 0.$$

Suppose that n such components are tested yielding failure times $t_1, t_2, \ldots,$ t_n. It is desired to estimate the mean failure time. First, the mean failure time is

$$\mu = E(T) = \int_0^\infty t \lambda e^{-\lambda t} \, dt = \frac{1}{\lambda}.$$

From the property noted above, the MLE of $1/\lambda$ would be given by $1/\hat{\lambda}$, where $\hat{\lambda}$ is the MLE of λ. Thus, we shall proceed to obtain the MLE of λ. The likelihood function is

$$L(\lambda) = \prod_{i=1}^{n} \lambda e^{-\lambda t_i} = \lambda^n e^{-\lambda \Sigma t_i},$$

and the log-likelihood function is

$$\ln L(\lambda) = n \ln \lambda - \lambda \sum t_i.$$

Setting

$$\frac{\partial \ln L(\lambda)}{\partial \lambda} = \frac{n}{\lambda} - \sum t_i = 0$$

and solving for λ yields $\hat{\lambda} = n/\sum t_i = 1/\bar{t}$. It follows that the MLE of the mean failure time is $\bar{t}$ which is the sample mean failure time. Therefore, the desired maximum likelihood estimate of the mean failure time for the ten given observations is calculated as 1.4. ▲

5.4 Basic Principles of Confidence Interval

As remarked in Sec. 5.2, a point estimate is a random variable distributed in some way around the true value of the parameter, and it is quite unlikely that the estimate is exactly equal to the true parameter value. Moreover, point estimates do not provide means of assessing the reliability or confidence we can place on them. Thus, it may be desirable to have an interval which we are fairly confident will include the true value θ. Such an interval is called a confidence interval. More precisely, the $100(1 - \alpha)\%$ *confidence interval* for θ is given by lower and upper limits θ_L and θ_U which satisfy

$$P(\theta_L \leq \theta \leq \theta_U) = 1 - \alpha. \tag{5.2}$$

Usually, α is taken to be small; for example, $\alpha = 0.05$ or $\alpha = 0.01$. If $\alpha = 0.05$, for example, then the interval (θ_L, θ_U) is called the 95% confidence interval.

The limits θ_L and θ_U are obtained from a sample, and they are random variables. Since θ is a fixed value although unknown, once a confidence interval is calculated, it either does or does not contain the true value of θ. If we were to take a large number of samples from the distribution, and if for each of the samples were to obtain $100(1 - \alpha)\%$ confidence interval for θ, then $100(1 - \alpha)\%$ of these confidence intervals would contain θ. In this sense we are $100(1 - \alpha)\%$ confident that a computed confidence interval covers the true value of θ.

The general method for obtaining confidence intervals is beyond the scope of this text. However, we shall illustrate a confidence interval using a simple example. Suppose that $X_1, X_2, \ldots, X_n$ are a sample from the normal distribution $N(\mu, \sigma^2)$ where σ^2 is known. First, we know that $\bar{X}$ has the normal distribution $N(\mu, \sigma^2/n)$. This means that the variable Z,

$$Z = \frac{\bar{X} - \mu}{\sigma/\sqrt{n}}$$

has the standard normal distribution. Denoting the $100\alpha/2\%$ point of the standard normal distribution by $z_{\alpha/2}$, we can write

$$P\left(-z_{\alpha/2} \le \frac{\bar{X} - \mu}{\sigma/\sqrt{n}} \le z_{\alpha/2}\right) = 1 - \alpha$$

Rearranging the inequalities inside the parentheses, we can rewrite

$$P\left(\bar{X} - z_{\alpha/2}\frac{\sigma}{\sqrt{n}} \le \mu \le \bar{X} + z_{\alpha/2}\frac{\sigma}{\sqrt{n}}\right) = 1 - \alpha. \tag{5.3}$$

Equation (5.3) is in the form of (5.2), and $[\bar{X} - z_{\alpha/2}(\sigma/\sqrt{n}), \bar{X} + z_{\alpha/2}(\sigma/\sqrt{n})]$ is the $100(1 - \alpha)\%$ confidence interval for the unknown μ.

The reader should be aware of the very serious error when the confidence interval is wrongly interpreted and misused as an interval which contains a large proportion of a population. The only interpretation is that a confidence interval is likely to contain the parameter value, for instance, the mean in the above example. In this regard, notice that the length of the confidence interval approaches zero as n increases.

Example 5.4 A sample of size 4 is taken from a normal distribution $N(\mu, 0.16)$. The observations were 7.6, 8.4, 7.9, and 8.1. To determine the 95% confidence interval for μ, we note that

$$1 - \alpha = 0.95, \quad \alpha/2 = 0.025, \quad z_{\alpha/2} = 1.96, \quad \bar{X} = 32.0/4 = 8.0,$$
$$\sigma/\sqrt{4} = 0.4/2 = 0.2.$$

From (5.3) the 95% confidence interval is given by

$$8.0 - 1.96(0.2) \quad \text{to} \quad 8.0 + 1.96(0.2),$$

that is, 7.6 to 8.4.   ▲

5.5 Basic Ideas of Hypothesis Testing

We shall now discuss one of the most important concepts in statistics, that of testing hypotheses. For instance, if a sample is known to have come from a normal distribution, is it reasonable that it could have come from one having a specified mean μ_0? If two samples are obtained from normal distributions, is it reasonable to state that they have equal means? A *statistical hypothesis*, or simply a *hypothesis*, is an assumption or statement about the population under study. Since the population is specified by one or more parameters of its distribution, a hypothesis is an assumption or statement regarding one or more parameters of a population distribution. The following are examples of hypotheses: (a) The average mileage per gallon of a particular car is 15 miles. (b) A particular diagnostic technique is 90% accurate. (c) The average lifetime of two different sets of electric bulbs is the same.

A procedure or decision rule which leads to acceptance or rejection of a hypothesis is called a *statistical test*. The hypothesis being tested is often represented by H_0 called the *null hypothesis* along with a mathematical specification about the parameter under H_0. It is also customary to specify, or at least to bear in mind, the *alternative hypothesis* denoted by H_1 which will be accepted in case H_0 is rejected. As an illustration, consider the above three examples of hypotheses. Suppose we denote the average mileage by μ, the proportion of correct diagnoses by p, and the average lifetime of the two different sets of bulbs by μ_1 and μ_2. Then three sets of null and alternative hypotheses can be:

$$\text{(a)} \quad H_0: \mu = 15, \qquad H_1: \mu \neq 15,$$

$$\text{(b)} \quad H_0: p = 0.9, \qquad H_1: p < 0.9,$$

$$\text{(c)} \quad H_0: \mu_1 = \mu_2, \qquad H_1: \mu_1 \neq \mu_2.$$

The alternatives H_1 of form (a) and (c) are called *two-sided*, and that of form (b) is called *one-sided*. The problem as to whether an alternative should be formulated as being one-sided or two-sided will be discussed in Sec. 5.5.2.

5.5.1 Rationale of Tests

Before proceeding into a detailed discussion of specific statistical tests, it would be helpful to pose the question of how we reject or accept the stated hypothesis H_0. Basically, we reject H_0 if the data do not support the hypothesis and if we think the alternative hypothesis H_1 is more tenable. How do we decide that the data do not support H_0? We do that if the probability of the occurrence of the observed event, or a more extreme one, calculated from the data is so small that H_0 is not likely to be true. We shall look at the problem more formally.

A statistical test is based on information obtained from data in the form of a *test statistic* which is a (single valued) function of the observations and a criterion for rejecting H_0. The sample median, sample means, and sample variance are only a few examples of statistics. So is any function of these statistics so long as it does not contain any unknown parameter. The statistics such as these, when determined from samples, are themselves variables having their own distributions. Thus, we can calculate the probability that the test statistic falls in a certain region when H_0 is true. Based on the probability level chosen, we can determine a region such that if the test statistic falls in the region, then we reject H_0. Clearly, the region should be such that the probability of rejecting H_0 when it is true is small. This region of the distribution is called the *rejection region, or critical region*. The *acceptance region* for H_0 is given by the complement of the critical region. Thus, for example, if $\bar{X} < 14.5$ is a critical region for a certain hypothesis H_0, then $\bar{X} \geq 14.5$ becomes the acceptance region.

5.5.2 One-Sided and Two-Sided Tests

To illustrate the idea, suppose we are interested in testing the hypothesis that the mean mileage per gallon (MPG), μ, of a model of car driven at a certain specified condition is 15. This hypothesis can be formulated as H_0: $\mu = 15$. In testing the H_0, suppose we are concerned with the possibilities that the mean MPG is above or below 15. The alternative is that the mean MPG does not equal 15, and it should be formulated as a two-sided alternative, H_1: $\mu \neq 15$.

Suppose that n identical cars of the given model are driven at the specified condition. It is recognized that there is a certain amount of variability in the mileage obtained so that the MPG is not a constant. Let $\bar{X}$ denote the sample mean MPG. We shall decide to use $\bar{X}$ as a test statistic for testing the above hypothesis H_0 without giving a very convincing reason at the moment. However, it is intuitively a reasonable variable to use for the purpose, in light of the fact that $\bar{X}$ is an estimator of μ. Under the given formulation of the H_1, it is intuitively clear that the evidence would not support the H_0, and the H_1 is more tenable if $\bar{X}$ is sufficiently different from 15. An example of a test is: reject the hypothesis H_0 that $\mu = 15$ if $\bar{X} < 14$ or if $\bar{X} > 16$; otherwise, accept it. The critical region in this case is "two-sided," and the test is referred to as a *two-sided test*.

Next, consider the circumstance that we are concerned only when the MPG is below 15 miles, but not if it is above that figure. In this case, it is appropriate to formulate the alternative in the form H_1: $\mu < 15$. Under this formulation an intuition would suggest that we reject H_0 if $\bar{X}$ is sufficiently smaller than 15. An example of a test is: reject the H_0 if $X < 14.5$; otherwise, accept it. The critical region in this case is given by $\bar{X} < 14.5$ which is "one-sided," and the test is referred to as a *one-sided test*.

Finally, a somewhat less likely situation in the present example is that we are concerned only when the mean MPG is over 15, but not if it is really below 15. In this case, the correct formulation of the alternative is H_1: $\mu > 15$, and we would use a one-sided test. For example, we decide to reject the H_0 if $\bar{X} > 15.5$, since the evidence would not support the H_0 if $\bar{X}$ is sufficiently larger than 15.

The above example shows that whether we use a one-sided or two-sided test depends on the alternative we have in mind. The decision as to whether the problem should be formulated as a one-sided or two-sided alternative depends upon the particular situation as well as on the question to be answered. The basic idea is that the one-sided alternative is appropriate in any experiment if the investigator is concerned only a priori with the difference in one direction. To be precise, the one-sided formulation would be justified:

(a) if the difference in only one direction (for example, $\mu_1 > \mu_2$) is important, whereas the other direction (for example, $\mu_1 < \mu_2$) is not, and

(b) if the investigator before the experiment is absolutely sure that the difference, if any, is going to be in one specified direction because of the nature of the response.

In all other circumstances, the problem should be set up in the two-sided alternative. We shall see that the two-sided test is more conservative in the sense that we are less likely to reject H_0 when it is true. Therefore, in case of doubt, it is suggested that we formulate H_1 as being two-sided. It is cautioned that the one-sided test is not justified just because the difference after completion of the experiment is found to be very large in one direction.

It is worthwhile to remember that the two-sided test is appropriate when we are concerned about the falsity of the null hypothesis without regard to the direction of the falsity.

5.5.3 Risk Involved in Decision Based on Tests

The previous section left much ambiguity and arbitrariness as to the selection of a test statistic and the formulation of the critical region. The given procedures were examples taken from an infinite number of such procedures. In other words, we could have used different statistics or different critical regions or both. It is evident that we wish to choose the procedure which leads to the correct conclusion most of the time. However, since the procedure depends on a test statistic which is a variable, it is rather impossible to hope for the correct decision always.

Whenever a decision or conclusion is made from a statistical test, there is a risk of committing an error. For example, a decision may be to reject H_0 when it is true. This error is defined as a *Type I error, and* the probability of a Type I error is denoted by α. A *Type II error* is an error of accepting H_0 when it is false, and its probability is denoted by β. Thus, α and β are conditional probabilities, which can be expressed as:

$$\alpha = P(\text{reject } H_0 \,|\, H_0 \text{ is true}),$$

$$\beta = P(\text{accept } H_0 \,|\, H_1 \text{ is true}).$$

The probability of correctly rejecting H_0 when H_1 is true is known as the power of the test. In other words, the power of the test denoted by π is given by

$$\pi = 1 - \beta.$$

Schematically, the various possibilities of outcomes involved in a statistical test can be summarized in Table 5.2.

Consider the problem of testing a hypothesis H_0 on the basis of a sample of size n. The first problem, which is not discussed at the moment, of course, is to determine a test statistic to be used. The next important question is

TABLE 5.2 Outcomes involved in statistical decisions

True situation

Our decision	H_0 true	H_1 true
Accept H_0	Correct decision	Type II error
Reject H_0	Type I error	Correct decision

how shall we determine the "optimal" critical region. Clearly, it would seem that the optimal critical region will be the one which minimizes α and β simultaneously. Unfortunately, it turns out that such a critical region does not exist. Usually, if we try to minimize α, then β is increased, and if β is minimized, then α gets larger. For example, consider a test which always accepts H_0 regardless of the outcome of observations. Clearly, α is equal to zero, but β could be large. On the other hand, if we decide unconditionally to reject H_0, then β will be equal to zero, but now α could become large.

An optimal critical region is formulated as follows. First, we fix α to a preassigned value depending on the Type I error that can be tolerated: common choices for α are 0.1, 0.05, 0.01, or other small values. After fixing α, the optimal critical region is given by the critical region that minimizes β. The idea is discussed in Sec. 5.7.

The numerical value selected for α is called the *significance level*, or α *level*, of the test. As stated already, the selection of a significance level depends on the Type I error that can be tolerated or the attitude toward the null hypothesis H_0. If we do not want to reject the hypothesis without extremely convincing evidence, then we should accordingly set the significance level very low. Otherwise, we wish not to set it too low. Another consideration can be the problem of trade-off between the significance level and the β value. As has been remarked already, if we set the significance level too low, the β value may become too large.

In applications, it is a good idea to determine not merely whether the given hypothesis is accepted or rejected at the specified significance level, but also to determine the smallest significance level at which the hypothesis can be rejected for the given evidence. This number enables others to make a decision based on the significance level of their own choice. The idea and concept are described in Sec. 5.8 in some detail.

In practice, as we will see, the test is performed with only the significance level α fixed. In this regard, an important aspect to remember is this: when a hypothesis H_0 is rejected, using a small α level, the resulting error will be small since the error involved is controlled by the α level. The smaller the α level, when the H_0 is rejected, the more convinced we are that the hypothesis is false. On the other hand, if H_0 is accepted, the error is unknown unless the value of β is calculated, which is usually not done and is often difficult to do.

It must be borne in mind that the acceptance of H_0 alone does not necessarily imply the truth of H_0, but simply the lack of evidence to reject H_0 at the given significance level. The procedure of testing the hypothesis is analogous to that applied in trying to convict in a court a person accused of a crime who is claiming innocence. In such trials, the hypothesis is that the person is innocent. Acquittal by the court does not necessarily mean the innocence of the accused, but simply the lack of evidence to convict him.

This section dealt with the general concept involved in testing the hypothesis. The concept will be illustrated next using the problem of a test for the mean of a normal distribution having a specified value, assuming that the standard deviation is known.

5.6 Tests Concerning Mean of Normal Distribution with Known Variance

Let $X_1, X_2, \ldots, X_n$ be a random sample of size n from a normal distribution with unknown mean μ and known variance σ^2. Suppose we wish to test a hypothesis which states that the mean is equal to some specified value, μ_0. In testing the hypothesis, suppose we are concerned only if the true mean μ is greater than μ_0, but not if μ is less than μ_0. The desired hypothesis and alternative can be formulated as:

$$H_0: \mu = \mu_0, \qquad H_1: \mu > \mu_0.$$

Let us assume that the significance level is chosen to be α. Let $\bar{X}$ denote the sample mean calculated from the sample. We know that this sample mean is also normally distributed with mean μ and variance σ^2/n.

It is intuitively clear that the evidence would not support the hypothesis H_0, and that H_1 is more tenable if $\bar{X}$ is sufficiently large relative to μ_0. In other words, it would seem reasonable to reject H_0 if $\bar{X}$ is sufficiently large relative to μ_0. The idea provides an intuitive rationale for the use of $\bar{X}$ as a test statistic and for a test procedure. It suggests that we should reject H_0 if $\bar{X} > c$, where c is a fixed constant yet to be determined, and accept H_0 if $\bar{X} \leq c$. Clearly, the constant c has to be such that the probability of $\bar{X}$ is greater than c would be small when the H_0 is true, since we want to reject the H_0 if $\bar{X} > c$. To be precise, the probability must be equal to α. To determine c, using the definition of α, we have:

$$\alpha = P(\text{reject } H_0 \,|\, H_0 \text{ is true})$$
$$= P(\bar{X} > c \,|\, \mu = \mu_0)$$
$$= P\left(\frac{\bar{X} - \mu_0}{\sigma/\sqrt{n}} > \frac{c - \mu_0}{\sigma/\sqrt{n}}\right). \tag{5.4}$$

Since $(\bar{X} - \mu_0)/(\sigma/\sqrt{n})$ has the standard normal distribution if H_0 is true,

(5.4) can be written as

$$\alpha = P\left(Z > \frac{c - \mu_0}{\sigma/\sqrt{n}}\right) \tag{5.5}$$

where Z has $N(0, 1)$. Equation (5.5) will be satisfied if we set

$$\frac{c - \mu_0}{\sigma/\sqrt{n}} = z_\alpha,$$

or

$$c = \mu_0 + z_\alpha \frac{\sigma}{\sqrt{n}}$$

where z_α is the upper $100\alpha\%$ point of the standard normal distribution. Thus, for a specified α, the critical region is given by

$$\frac{\bar{X} - \mu_0}{\sigma/\sqrt{n}} > z_\alpha$$

or, equivalently,

$$\bar{X} > \mu_0 + z_\alpha \frac{\sigma}{\sqrt{n}}. \tag{5.6}$$

For example, if $\alpha = 0.05$, then we reject H_0 if $\bar{X} > \mu_0 + 1.645\,\sigma/\sqrt{n}$ and accept H_0 otherwise.

Summarizing, the test of the hypothesis $H_0: \mu = \mu_0$ against $H_1: \mu > \mu_0$ when σ^2 is known can be performed by rejecting H_0 at level α if

$$Z = \frac{\bar{X} - \mu_0}{\sigma/\sqrt{n}} > z_\alpha.$$

This test makes sense because a large value of Z provides support for H_1 and not H_0. The test is one-sided.

Suppose we wish to test the same hypothesis H_0 against the alternative of the form

$$H_1: \mu < \mu_0.$$

Using an argument similar to that made above, the rejection region of the test is obtained as

$$Z = \frac{\bar{X} - \mu_0}{\sigma/\sqrt{n}} < -z_\alpha.$$

Now suppose we wish to test the hypothesis H_0 against the two-sided alternative

$$H_1: \mu \neq \mu_0.$$

In this case, both large and small values of $\bar{X}$, and hence Z, provide support for H_1. Thus, it is reasonable to let both large and small values of Z comprise the rejection region. For a test with significance level α, since the value of α must be allocated to each side, the hypothesis H_0 is rejected if

$$Z = \frac{\bar{X} - \mu_0}{\sigma/\sqrt{n}} < -z_{\alpha/2} \quad \text{or} \quad Z > z_{\alpha/2}.$$

The rejection regions for testing the hypothesis $H_0: \mu = \mu_0$ at $\alpha = 0.05$ level depending on the three different forms of the alternative H_1 are sketched in Fig. 5.3.

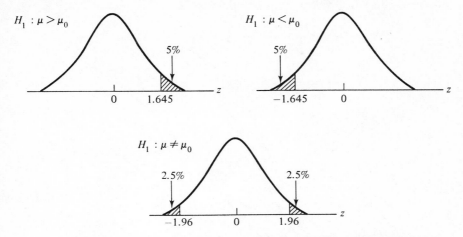

Fig. 5.3   Rejection regions for testing $H_0: \mu = \mu_0$ when σ is known

The curves shown in Fig. 5.3 are graphs of the standard normal distribution. Since the variable $Z = (\bar{X} - \mu_0)/(\sigma/\sqrt{n})$ has the standard normal distribution if $\mu = \mu_0$, the graphs are the exact shape of the distribution of the variable when H_0 is true. As a consequence, if the H_0 is true and if the given rejection regions are used, the chance of an erroneous decision to reject the H_0 will be 5% in each case.

As previously mentioned, the evaluation of the probability of a Type II error, β, is important in assessing the efficiency of any test. We shall proceed to calculate β for the test given by (5.6). In order to do this, let us assume, to be more specific, that $\mu = \mu_1$ where $\mu_1 > \mu_0$ when H_1 is true. The probability β can be determined as

$$\beta = P(\bar{X} \leq c \,|\, H_1 \text{ is true})$$

$$= P\left(\bar{X} \leq \mu_0 + z_\alpha \frac{\sigma}{\sqrt{n}} \,\middle|\, \mu = \mu_1\right)$$

$$= P\left(\frac{\bar{X} - \mu_1}{\sigma/\sqrt{n}} \leq \frac{\mu_0 - \mu_1 + z_\alpha(\sigma/\sqrt{n})}{\sigma/\sqrt{n}}\right)$$

$$= P\left(Z \leq z_\alpha - \frac{\mu_1 - \mu_0}{\sigma/\sqrt{n}}\right). \tag{5.7}$$

Suppose that $\alpha = 0.05$, $\mu_1 - \mu_0 = 1$, $\sigma = 5$, and $n = 25$, for example. Then, β is given by $P(Z \leq 1.645 - 1) = 0.74$, and the power of the test is $\pi = 0.26$. It is interesting and useful to note that the power can be made greater by increasing the sample size n for a fixed α, and a fixed difference $d = \mu_1 - \mu_0$. Similarly, for a fixed n it is evident that the power increases as the difference $d = \mu_1 - \mu_0$ increases. The graph of the power for several different values of n and d when $\alpha = 0.05$ and $\sigma = 5$ is sketched in Fig. 5.4.

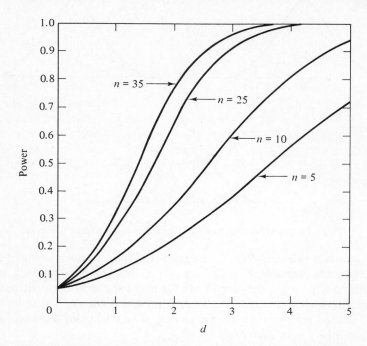

Fig. 5.4 Power for testing $H_0: \mu = \mu_0$ vs. $H_1: \mu > \mu_0$ as a function of n and $d = \mu_1 - \mu_0$ when $\sigma = 5$ and $\alpha = 0.05$

Example 5.5 A certain brand of bread is advertized as having an average of 70 calories per slice. Tests on a sample of 25 slices yield $\bar{X} = 71$ calories. If it is known that $\sigma = 4$ calories, what can be said about the claim? Assume $\alpha = 0.05$.

The hypothesis and alternative can be formulated as

$$H_0: \ \mu = 70, \qquad H_1: \ \mu \neq 70,$$

where μ denotes the mean calories per slice. We calculate

$$Z = \frac{71 - 70}{4/\sqrt{25}} = 1.25.$$

The two-sided test rejects the hypothesis H_0 at $\alpha = 0.05$ level if either $Z < -1.96$ or $Z > 1.96$; since $-1.96 < 1.25 < 1.96$, the test does not reject the H_0. Thus, the available evidence does not contradict the claim. Suppose we are not concerned about the lower calories but only about the calorie level greater than 70. The alternative should be formulated as $H_1 : \mu > 70$. On the other hand, if the only concern is that the bread has less than 70 calories per slice on the average, then the alternative should be $H_1 : \mu < 70$. ▲

*5.7 Power and Choice of Test

We shall elaborate further on some of the general concepts outlined in Sec. 5.5. Again, consider the problem of testing the hypothesis $H_0 : \mu = \mu_0$ against the alternative $H_1 : \mu > \mu_0$ regarding the mean of a normal distribution with known variance. To be more specific, let us assume that $\mu = \mu_1$ where $\mu_1 > \mu_0$ when H_1 is true. Let $g_0(\bar{x})$ denote the density function of $\bar{X}$ when H_0 is true and $g_1(\bar{x})$ when H_1 is true. Note that $g_0(\bar{x})$ is $N(\mu_0, \sigma^2/n)$ and $g_1(\bar{x})$ is $N(\mu_1, \sigma^2/n)$. We shall first consider the rejection region of the form $\bar{X} > c$ given by (5.6), where c is a constant which depends on α. The relation between $g_0(\bar{x})$, the rejection region, and α is illustrated in the top figure of Fig. 5.5. From the definition of α, the constant c is required to satisfy the equation

$$\alpha = P(\text{reject } H_0 \,|\, H_0 \text{ is true})$$

$$= P(\bar{X} > c \,|\, H_0 \text{ is true}) = \int_c^\infty g_0(\bar{x})\, d\bar{x}.$$

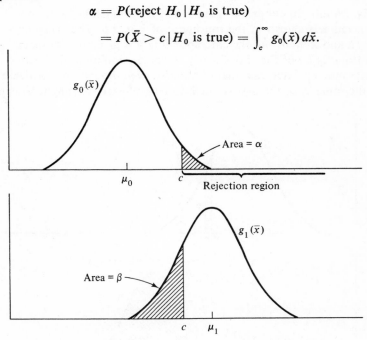

Fig. 5.5   Relation between a critical region, α and β

Consider what will happen if H_1 is true so that $g_1(\bar{x})$ is the true density function. Since the acceptance region is given by $\bar{X} \leq c$, β must satisfy the equation

$$\beta = P(\text{accept } H_0 \,|\, H_1 \text{ is true})$$

$$= P(\bar{X} \leq c \,|\, H_1 \text{ is true}) = \int_{-\infty}^{c} g_1(\bar{x}) \, d\bar{x}.$$

The situation is depicted in the bottom figure of Fig. 5.5 where β is given by the shaded area. From Fig. 5.5 it can be seen that if we move c to the right, α will become smaller but β becomes larger, and if c is moved to the left, β will become smaller but now α becomes larger. In other words, a decrease in α causes an increase in β and vice versa.

It should be noted that an infinite number of rejection regions are, in general, possible for a given H_0 and a given α. In the present problem, we could, for example, construct a rejection region of the form $a < \bar{X} < b$, where a and b are constants. All we have to require for a and b are that they satisfy

$$\alpha = P(a < \bar{X} < b \,|\, H_0 \text{ is true})$$

$$= \int_{a}^{b} g_0(\bar{x}) \, d\bar{x}.$$

Clearly, an infinite number of pairs a and b are possible, and one such pair for a given α is shown in the top figure of Fig. 5.6. The acceptance region is $\bar{X} \leq a$ and $\bar{X} \geq b$, and the corresponding β is given by the shaded area of the bottom figure of Fig. 5.6. Clearly, β corresponding to the second critical region is much greater than that of the first, even though both critical regions have the same α, so the first is much to be preferred. As might be expected,

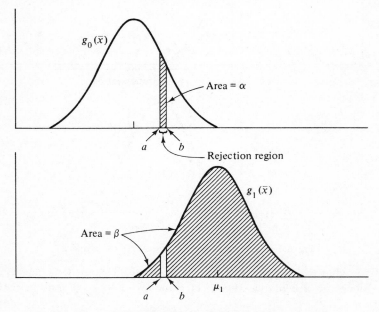

Fig. 5.6   Relation between a bad critical region, α and β

the first critical region illustrated in Fig. 5.5 can be proven to be the optimal for testing $H_0: \mu = \mu_0$ against $H_1: \mu > \mu_0$.

The obvious basis for comparing any two tests having the same Type I error probability is the relative sizes of their Type II error probabilities. A test is said to be more *powerful* or more *efficient* if its Type II error probability is smaller than that of the other test.

5.8 *P* Value

The basic problem of testing a hypothesis H_0 is to find a rejection region with the significance level α where the value of α is usually chosen in advance. The commonly used values of α are 0.05 and 0.01, although arbitrary. Then, we decide to reject or accept the hypothesis depending on whether the test statistic falls in the rejection region or not. Suppose that a hypothesis can be rejected on the basis of a significance level of 0.05. Then, we state, for example, "we reject the hypothesis at 5% level" or "the mean is significantly different from μ_0 at 5% level." It is conceivable that the result of the analysis is such that the hypothesis can be rejected at a significance level smaller than 0.05. There is no doubt that the statements such as the above do not provide the precise information as to the test result: they do not give an idea of how strongly the data contradict (or support) the hypothesis.

In Sec. 5.5, we discussed the idea of determining the smallest significance level at which the hypothesis can be rejected. For a given observation and a hypothesis, the smallest level at which the H_0 could be rejected is known as *P value*. The *P* value can be interpreted as the probability of a sample outcome equal to or more extreme than that observed when the hypothesis H_0 is true. Thus, in a sense, it measures a degree of disagreement between the observed value of the test statistic and its expected value under the H_0. Small values of P may be interpreted as evidence against the hypothesis, although the magnitude of the evidence depends on the efficiency of the test which was briefly discussed in Sec. 5.7.

As an illustration, suppose we wish to test the following hypothesis H_0 against the alternative H_1 about the mean of a normal distribution with the known σ of 1.0:

$$H_0: \mu = 8.0, \qquad H_1: \mu < 8.0.$$

Suppose that a sample of $n = 9$ gave $\bar{X} = 7.0$. Then, the value of Z is calculated to be

$$Z = \frac{7.0 - 8.0}{1.0/\sqrt{9}} = -3.0.$$

The hypothesis H_0 can be rejected at the significance level of $\alpha = 0.05$. However, we know that H_0 can be rejected at a level smaller than 0.002 since $P(Z < -3.0)$ is less than 0.002. To be precise, P value in this example is given by $P(Z < -3.0) = 0.0013$. In the example, if the alternative is two-sided, that is, if $H_1: \mu \neq 8.0$, the critical region is located in both tails of the distribution, and P value is given by doubling the one-sided P value: in this example, $P = 0.0026$. On the other hand, if $H_1: \mu > 8.0$, we have $P = P(Z > -3) > 0.05$, and the proper decision is not to reject the hypothesis H_0.

In Fig. 5.7, the shaded area gives the precise value of P, namely, 0.0026, under the two-sided alternative. In practice, however, there is no need for obtaining the exact P value, but simply to determine if $P < 0.05$, $P < 0.01$, $P < 0.005$, or $P < 0.001$, or some other convenient level. One way of summarizing the result of an analysis is to state the conclusion and the P value simultaneously with the format: "the difference is statistically significant at $P < 0.01$" or "not statistically significant $(P > 0.05)$." Often we simply state "not statistically significant" without indicating the P value if it is greater than a predetermined α level.

In reporting the result of any statistical analysis, giving P value like 0.001 is considerably more informative than a bare-bones statement such as "reject H_0 at level 0.05" or "significant at level 0.05."

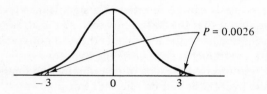

Fig. 5.7 P value for two-sided test

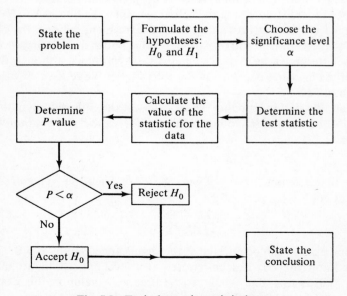

Fig. 5.8 Typical steps in statistical test

It must be emphasized that a small P value does not necessarily indicate a correspondingly large probability of the truth of the alternative H_1. Remember that the P value is calculated under H_0, and H_1 does not enter into the calculation. Neither does the P value indicate the probability of the truth of the hypothesis. Further, as mentioned at the end of Sec. 5.5, it is important to remember again that "not significant" does not prove the hypothesis H_0 to be true. The most that can be said with results that are not significant is that there is not sufficient evidence to contradict the hypothesis based on the information obtained from the data.

We conclude this section with Fig. 5.8, which illustrates a suggested procedure for testing a hypothesis based on real data. It would be a good idea, at least, to keep in mind the steps similar to those shown in the flow chart in testing a hypothesis.

5.9 Problem of Sample Size

In planning a statistical study, a frequent question regards the size of the sample. The question is important because too large a sample could lead to a waste of resources, while too small a sample could diminish the utility of the results. The answer to the question of the required sample size depends roughly on criteria involving the precision desired with the result, and the amount of risk that can be tolerated or the confidence that can be placed on the result.

In particular, consider an estimation problem. The criterion involves the degree of accuracy required for the estimate, and the variance or standard deviation of the estimator can be used as a measure of accuracy. An alternate method might be to specify the probability with which the estimate falls within a given distance of the quantity being estimated. A simple example may make the general idea clear.

As the first example, suppose it is desired to estimate the proportion p of a population possessing a certain characteristic. The criteria may be specified as follows: we like to be $100(1 - \alpha)\%$ confident that the estimated proportion is correct within $\pm\delta$ of the true proportion p, for a small value of δ. The required sample size depends on the values of α and δ. For the second example, consider the problem of testing hypotheses $H_0: \mu = \mu_0$ and $H_1: \mu = \mu_1$ about the mean of a normal distribution. Since two errors in a decision procedure are Type I and Type II errors, the desired sample size will depend on the values of α and β.

We shall illustrate the method of approximating the sample size n for the first example. The sample size problem for the second example will be discussed in Sec. 6.9. In the first example, we are concerned with a problem of estimating p. Let X be the number of observations with the characteristic in a sample of n observations. Consider $\hat{p} = X/n$ which is an estimator for p. Due to the Central Limit Theorem, $\hat{p}$ has an approximate normal distribution for a large n with its variance given by pq/n, applying (2.23) with $a = 1/n$ to (3.3). Since $\hat{p}$ can be assumed to be normally distributed with the mean p, it will be in the range

$$p \pm z_{\alpha/2} \sqrt{pq/n},$$

with the probability equal to $1 - \alpha$. Because δ is the margin of error in the estimation, we can set

$$z_{\alpha/2} \sqrt{pq/n} = \delta.$$

Solving for n yields a formula:

$$n = \frac{z_{\alpha/2}^2 pq}{\delta^2}. \qquad (5.8)$$

At this point, a rather awkward difficulty is noted which is common to all problems in the determination of sample size. That is, the formula for n involves some unknown parameters of the population. In this instance, (5.8) depends on the value of p. However, the formula can still be used to determine n if some a priori knowledge about p is available. As an illustration, suppose that p is known to lie between 0.1 to 0.4. Then, the product pq lies between 0.09 and 0.24, and the sample size will be maximum when $pq = 0.24$. To be on the safe side, $pq = 0.24$ can be substituted in (5.8) to determine the required n. If nothing is known about p or in case of doubt, n can be approximated by

$$n = \frac{z_{\alpha/2}^2}{4\delta^2}, \qquad (5.9)$$

since 0.25 is the absolute maximum of pq.

Example 5.6 Suppose it is desired to estimate the proportion p of a population favoring a candidate for the presidency and we want to be 95% confident that the estimate is correct within 2.5%.

Substituting $z_{0.025} = 1.96$ and $\delta = 0.025$ into (5.9) we find the sample size of $n = 1537$. It is interesting to remark that this is about the sample size used by many reputable pollsters in public opinion polls. ▲

It is worthwhile to note the following properties about Eq. (5.9): (a) First, as might be expected, a larger sample size is required if the margin of error δ is reduced for a fixed level of confidence or if the confidence level is increased for a fixed δ, or when δ is made smaller and the confidence level is made larger. (b) Secondly, the formula, strictly speaking, is valid for an infinite population. However, it is adequate for a population of 100,000 or one billion as long as it is very large.

PROBLEMS

1. Two teams, A and B, were sent to the same forest to take a random sample of 100 trees each and to measure the circumferences of the trees. The results in inches were: $\bar{X}_A = 14.3$ and $s_A = 1.8$ by A, and $\bar{X}_B = 13.2$ and $s_B = 1.6$ by B. (a) Does the smaller standard deviation for B imply that the result by B is more accurate? (b) If it is known that the true mean is 13.9, which sample mean is more accurate? (c) Which sample standard deviation is more accurate in this case?

*2. Suppose that the time required by workers to assemble a certain part is approximately normally distributed with the variance of 20 min. (a) If the mean assembly time is estimated from a sample of 20 workers, what is the mean square error (MSE) of the estimate? (b) What is the MSE if the sample size is 40?

3. A sample of 16 cigarettes of a certain brand was tested for nicotine content and gave $\bar{X} = 1.2$ mg. Find the 99% confidence interval for the mean μ given that $\sigma = 0.2$ mg.

4. Calculate the 90% confidence interval for the mean μ of a normal distribution if $\sigma = 2$ and when a sample of size 8 gave the values 10, 8, 11, 13, 9, 10, 7, 11.

5. A group of 100 girls randomly selected from a university has an average weight of 125 lb. (a) Determine the 95% confidence interval for the mean weight given that the standard deviation is 20 lb. (b) Interpret the confidence interval.

*6. Out of 40 identical thumbtacks tossed, 28 are found to have landed with points up. Using a normal approximation, determine the 95% confidence interval for the true probability that a tack lands with point up.

*7. Suppose n is large enough to justify a normal approximation to the binomial distribution. Show that a 95% confidence interval for the proportion p is given by $p_1 < p < p_2$ where p_1 and p_2 are the solutions of the quadratic equation $(p - p^2)(1.96)^2 = n(\bar{X} - p)^2$ where $\bar{X}$ represents the sample fraction.

*8. Find the maximum likelihood estimate of q for $f(x) = pq^x$; $p = 1 - q$, $x = 1, 2, \ldots$, if n observations $X_1, X_2, \ldots, X_n$ are obtained.

*9. Find the maximum likelihood estimate of the parameter λ of a Poisson distribution if seven observations are 20, 17, 14, 18, 19, 18, 20.

*10. Determine the maximum likelihood estimate of the mean μ of a normal distribution if σ^2 is known.

*11. Find the joint maximum likelihood estimate of μ and σ^2 of a normal distribution, $N(\mu, \sigma^2)$.

12. In a court case in which a man is being tried for murder, what are the two types of error? Discuss the analogy between the two errors and Type I and Type II errors in testing a hypothesis.

13. A car is advertised as having gas mileage of 14 miles per gallon in high-way driving. The miles per gallon obtained in nine independent experiments were: 12, 14, 11, 12, 13, 13, 15, 11, 12. (a) Is the advertisement justified at $\alpha = 0.05$? (b) Determine the P value. Assume that $\sigma = 1.2$.

14. In a normal distribution with $\sigma = 4$, it is desired to test the hypothesis $H_0: \mu = 10$ against $H_1: \mu > 10$ with $\alpha = 0.05$. (a) With a sample of 49, what is the chance of rejecting H_0 if the true $\mu = 12$? (b) If $\mu = 14$? (c) If $\mu = 18$?

15. Repeat Prob. 14 when the sample size is 25. Compare the answers with those for Prob. 14.

*16. Referring to Prob. 14, how large a sample should be used if it is desired to have a chance of 0.90 of rejecting H_0 if $\mu = 12$? If $\mu = 14$?

17. A sample of 100 from a normal distribution with $\sigma = 2.5$ has been taken in order to test the hypothesis $H_0: \mu = 23.0$ against the alternative $H_1: \mu \neq 23.0$. If the significance level was set to $\alpha = 0.05$, what is the chance of falsely accepting the H_0 if the true mean happens to be $\mu = 22.0$?

*18. A standard drug is known to be effective in about 80% of cases in which it is used to treat infections. A new drug has been found effective in 85 of the first 100 cases tried. Is the superiority of the new drug well established? (If the new drug were equally effective as the old, what would be the probability of obtaining 85 or more successes in a sample of 100?)

*19. Suppose that a population has the following distribution

$$f(x) = \lambda e^{-\lambda x} \quad x \geq 0.$$

A reasonable critical region for testing $H_0: \lambda = 0.5$ against $H_1: \lambda = 1$ based on one observation is $X < c$ where c is a constant to be determined. (a) For $\alpha = 0.05$, determine the c. (b) Determine the power of this test.

*20. The government samples about 50,000 households to determine the unemployment rate. Suppose it is known that the rate is between 5% to 9% and that it is desired to have a 95% confidence about the result. What is the margin of possible error of the estimated rate?

*21. A research worker wishes to estimate the mean of a population using a sample large enough so the chance will be 95% that the sample mean will not differ from the population mean by more than 25% of the standard deviation. How large a sample should he take? Assume that the population has a normal distribution.

6

Estimation and Testing Hypotheses:
Frequency Data

6.1 Introduction

Situations frequently arise where observations are made on qualitative or nominal scales or in categories. Examples of data of this type are blood groups, the number of heart attacks per day in a certain city, the number of defective items in a lot, or enumeration of eye color. The basic characteristic of such data is that the data are obtained by counting the number of observations in a sample which fall into certain specified categories. Data obtained by counting the frequency in categories are called *frequency data* or *categorized data*. Some frequency data may be the result of a convenient classification of observations which can be measured. For example, we may be dealing with the frequencies of individuals with high and normal blood pressure in a given population. Men may be classified as either obese or normal in terms of their weight. On the other hand, some variables are inherently of categorized form. The blood groups, the number of cancer patients, or the number of breakdowns of a machine are examples of the variables which cannot be measured on a continuous scale.

In this chapter, we shall consider some useful statistical methods appropriate for analyzing frequency data. For example, we may wish to test the hypothesis that the proportion of good fuses is 95% or we may want to compare the proportion of cure for a disease using two treatments, or we may be concerned with the possible relationship between eye and hair colors, and so on. The methods discussed in this chapter are very frequently used in applications. An important point to remember is that the data used in statistical analyses described in this chapter consist of the number of observations in a sample which fall into specified categories.

6.2 Inference about a Proportion

Two of the frequently encountered problems are that of testing hypotheses and that of determining confidence intervals for an unknown proportion p of individuals in a population having a certain characteristic. It is convenient to define "success" or "failure" depending on whether an observation has the characteristic. Let X be the number of successes in a sample of n independent observations in which the probability of success is the constant p. The distribution of the number of successes X is given by the binomial distribution whose frequency function is

$$f(x) = \binom{n}{x} p^x q^{n-x}, x = 0, 1, \ldots, n,$$

where $q = 1 - p$ and $\binom{n}{x} = \dfrac{n!}{x!\,(n-x)!}$. See Sec. 3.2.

The best estimator for p denoted by $\hat{p}$ is given by an intuitive estimator:

$$\hat{p} = \frac{X}{n}.$$

Equation (3.3) gave the variance of X as npq, and it follows from (2.23) that the variance of $\hat{p}$ is pq/n. Thus, the standard deviation of X can be estimated as $\sqrt{n\hat{p}\hat{q}}$ and the standard deviation of $\hat{p}$ by $\sqrt{\hat{p}\hat{q}/n}$ where $\hat{q} = 1 - \hat{p}$.

6.2.1 Testing the Hypothesis about a Proportion

Consider the problems dealing with the proportion of defectives, the chance of cure or the chance of making money, and so on. We may wish to test a hypothesis H_0 that the defective proportion (or the proportion of cure) p equals a specified proportion p_0.

The basic statistic used to test hypotheses about p is X/n, the frequency of successes in a sample of n independent observations. If p_0 is the true proportion, using the Central Limit Theorem, for large n the variable

$$Z = \frac{\dfrac{X}{n} - p_0}{\sqrt{p_0 q_0/n}} \tag{6.1}$$

where $q_0 = 1 - p_0$ is known to have approximately a standard normal distribution. This fact enables us to construct the following tests using exactly the same rationale and method described in Sec. 5.6.

(a) $H_0: p = p_0$ vs. $H_1: p > p_0$.

  Clearly, H_0 should be rejected if X/n is large relative to p_0, or equivalently, if the value of Z is too large, since this would indicate that H_0 is likely to be false. For a test with the significance level α, the

H_0 is rejected if

$$Z = \frac{\dfrac{X}{n} - p_0}{\sqrt{p_0 q_0 / n}} > z_\alpha$$

where z_α is the upper $100\alpha\%$ point of the standard normal density function.

(b) $H_0 : p = p_0$ vs. $H_1 : p < p_0$.
We reject if Z is too small, to be precise, if

$$Z < -z_\alpha.$$

(c) $H_0 : p = p_0$ vs. $H_1 : p \neq p_0$.
Against this alternative H_1, both small and large values of Z contradict H_0 and provide support for H_1. If the significance level is α, then we reject H_0 if either

$$Z < -z_{\alpha/2} \quad \text{or} \quad Z > z_{\alpha/2}.$$

(d) If n is not large so that if $np_0 < 5$ or $nq_0 < 5$, the above three tests are no longer adequate as the normal approximation for the binomial distribution is not very satisfactory. Let x be the actual observed number of "successes" in n observations. For the alternative H_1: $p > p_0$, the exact P value is given by the probability of observing x or more "successes" when H_0 is true. Thus,

$$P = P(X \geq x) = \sum_{k=x}^{n} \binom{n}{k} p_0^k q_0^{n-k}. \tag{6.2}$$

If H_1 is $p < p_0$, the P value is given by

$$P = P(X \leq x) = \sum_{k=0}^{x} \binom{n}{k} p_0^k q_0^{n-k}. \tag{6.3}$$

The calculation of P values for two-sided alternatives can be performed as follows: if x is greater than np_0, the two-tailed P value is given by doubling the P value calculated from (6.2). If x is less than or equal to np_0, the two-tailed P value is given by doubling the P value given by (6.3).

Example 6.1 Consider a shipment consisting of large units of a certain manufactured product. Suppose the lot is acceptable only if the proportion of defectives is at most 2%. An inspection of 300 randomly selected units yielded nine defectives. Should the lot be rejected at $\alpha = 0.05$ level of significance?

Let p denote the true proportion of defectives. The problem can be formulated by the following hypothesis and alternative:

$$H_0 : p = 0.02, \qquad H_1 : p > 0.02.$$

Substituting $\hat{p} = X/n = 0.03$ into (6.1), we calculate

$$Z = \frac{0.03 - 0.02}{\sqrt{(0.02)\dfrac{0.98}{300}}} = 1.24.$$

Since this value is less than 1.645, the upper 5% of the standard normal distribution, the hypothesis cannot be rejected at the 5% level of significance. The available evidence does not suggest the rejection of the lot. ▲

6.2.2 Confidence Interval for a Proportion

Again, making use of the fact that for large n the binomial distribution is approximated by a normal distribution, the variable $(\hat{p} - p)/\sqrt{pq/n}$ where $\hat{p} = X/n$ can be treated as if it has the standard normal distribution. Further, if n is large, it is adequate to approximate pq by $\hat{p}\hat{q}$ without introducing any appreciable error. For large n, therefore, it follows that the variable defined by

$$Z = \frac{\hat{p} - p}{\sqrt{\dfrac{\hat{p}\hat{q}}{n}}}$$

has approximately a standard normal distribution. Hence, we can write

$$P\left(-z_{\alpha/2} \leq \frac{\hat{p} - p}{\sqrt{\dfrac{\hat{p}\hat{q}}{n}}} \leq z_{\alpha/2}\right) = 1 - \alpha$$

which leads to the approximate $100(1 - \alpha)\%$ confidence interval:

$$\hat{p} \pm z_{\alpha/2}\sqrt{\frac{\hat{p}\hat{q}}{n}}. \tag{6.4}$$

The approximation is better the closer p is to 0.5 and the larger the n. If the sample size is not large, say $n < 20$, a more exact method is needed to obtain the confidence interval. Tables A7 and A8 of the Appendix give the 95% and 99% confidence intervals for p.

Example 6.2 A new drug is tried on 20 patients and 17 showed some improvement. Determine the 95% confidence interval for the true proportion p of improvement using the drug.

From Table A7 we find the confidence interval is 62% to 97%. Roughly speaking, the chance is 95% that this interval embraces the true proportion. If Eq. (6.4) is used, we obtain $0.85 \pm 1.96\sqrt{(0.85)(0.15)/20}$ or the 95% confidence interval is given by 69% to 100%. The result is not very close to the exact limits as p appears to be considerably greater than 0.5 and n is not very large. ▲

6.3 Comparison of Two Proportions

Suppose that samples of size n_1 and n_2 are obtained from two populations and a and c of n_1 and n_2 are found to possess a certain characteristic. The basic data and estimated proportions can be summarized as in the following table.

| | Having characteristic | | | Estimated |
	Yes	No	Total	proportion
Population 1	a	b	n_1	$\hat{p}_1 = a/n_1$
Population 2	c	d	n_2	$\hat{p}_2 = c/n_2$

Based on the above information, we may wish to compare the population proportions p_1 and p_2. The hypothesis can be formulated as follows:

$$H_0: p_1 = p_2.$$

If the hypothesis is true, the combined pooled estimator $\hat{p}$ of the common population proportion, p, is

$$\hat{p} = \frac{a + c}{n_1 + n_2}.$$

Consider the variance of $\hat{p}_1 - \hat{p}_2$. Using (2.24), it can be estimated as

$$\frac{\hat{p}\hat{q}}{n_1} + \frac{\hat{p}\hat{q}}{n_2} = \hat{p}\hat{q}\left(\frac{1}{n_1} + \frac{1}{n_2}\right)$$

where $\hat{q} = 1 - \hat{p}$. Therefore, under H_0, by the Central Limit Theorem as in Sec. 6.2.2, the variable

$$Z = \frac{\hat{p}_1 - \hat{p}_2}{\sqrt{\hat{p}\hat{q}\left(\dfrac{1}{n_1} + \dfrac{1}{n_2}\right)}} \tag{6.5}$$

has approximately a standard normal distribution if n_1 and n_2 are large. This result enables us to perform an approximate test for H_0 when the sample sizes are large.

(a) $H_0: p_1 = p_2$ vs. $H_1: p_1 > p_2$.
Against this alternative, the H_0 should be rejected if the value of Z calculated from Eq. (6.5) is too large since this would support H_1 instead of H_0. We reject if

$$Z > z_\alpha.$$

(b) $H_0: p_1 = p_2$ vs. $H_1: p_1 < p_2$.
The situation is exactly the reverse of the above, and we reject if Z is too small, that is, if

$$Z < -z_\alpha.$$

(c) $H_0: p_1 = p_2$ vs. $H_1: p_1 \neq p_2$.

This is the two-sided test, and the rejection region is $Z < -z_{\alpha/2}$ or $Z > z_{\alpha/2}$. However, the equivalent and more common method of testing this two-sided hypothesis is the chi-square test which is to be described in Sec. 6.5.

Example 6.3 In a study of criminal recidivism, the parole record indications of 48 alcoholics and 55 nonalcoholics were obtained. The indications refer to the subsequent arrests, convictions, etc. The data (from 1965 *American Journal of Psychiatry*, 122, p. 436) is summarized in the following table. Is the chance of the indications significantly higher in alcoholics than in nonalcoholics?

	n	Number with indications	% of indicated parolees
Alcoholics	48	34	70.8
Nonalcoholics	55	11	20.0

Let p_A and p_N represent the expected proportion of the indications for alcoholics and nonalcoholics, respectively. The hypotheses can be formulated as $H_0: p_A = p_N$ and $H_1: p_A > p_N$. We calculate $\hat{p} = 0.437$ and

$$Z = \frac{0.708 - 0.200}{\sqrt{0.246(\frac{1}{48} + \frac{1}{55})}} = 5.19.$$

$$\frac{34 + 11}{48 + 55}$$

The proportion, p_A, is significantly greater than p_N at $P < 0.001$. We conclude that the subsequent crime rate by alcoholic parolees is significantly higher than that for nonalcoholic parolees. ▲

6.4 Goodness of Fit Test

A problem arising frequently in scientific studies is the testing of the *goodness of fit*, or the compatibility of a set of data with a theoretical frequency. For the first example, consider a genetic theory which states that under random mating an offspring belongs to the genotypes AA, Aa, and aa in the ratios $1:2:1$. Because of chance fluctuations, the actual observed genotype frequencies will almost never coincide exactly with the theoretical ratios. If a sample of 60 offspring under random mating yield 14, 28, and 18 with the three genotypes, respectively, we wish to test whether the data support the theory. For the second example, we may wish to test whether the incidence of heart attacks is the same in each of seven days of a week. Denoting the probability of the incidence on Sunday, Monday, etc., by p_1, p_2, etc., the hypothesis can be formulated as

$$H_0: p_1 = \tfrac{1}{7}, \quad p_2 = \tfrac{1}{7}, \ldots, p_7 = \tfrac{1}{7}.$$

Suppose the problem is to test whether the incidence is the same for the week days and week ends. Let p_1 be the probability of heart attack on week days and p_2 that on week ends now. Then, the problem can be hypothesized as

$$H_0: p_1 = \tfrac{5}{7}, \qquad p_2 = \tfrac{2}{7}.$$

Suppose an experiment can result in k mutually exclusive events or categories. Let p_i be the probability that the ith event will occur in a trial of the experiment. If p_{i0} denotes the postulated value of the probability, then the goodness of fit problem can be hypothesized as

$$H_0: p_i = p_{i0}, \qquad i = 1, 2, \ldots, k.$$

Let us assume that the total number of observations is n and that n_i of them fall in the ith category. If the hypothesis $H_0: p_i = p_{i0}$ is true, the expected frequencies are given by

$$e_i = np_{i0}.$$

Without giving the proof, we shall state the following theorem which provides a means of testing goodness of fit problems.

Theorem 6A

If $n_1, n_2, \ldots, n_k$ and $e_1, e_2, \ldots, e_k$ are the observed and expected frequencies, respectively, for k possible outcomes of an experiment, and if the total sample size n is large, the quantity

$$X = \sum_{i=1}^{k} \frac{(n_i - e_i)^2}{e_i} \tag{6.6}$$

has an approximate chi-square (χ^2) distribution with $k - 1$ degrees of freedom when H_0 is true.

The test based on this theorem is commonly known as the χ^2-*test of goodness of fit*. As we shall see, Theorem 6A provides the basis for various tests dealing with frequency data in addition to the goodness of fit problem.

We reject H_0 if the value of X calculated from (6.6) is larger than the critical value read from Table A3 at a given significance level. The χ^2-test

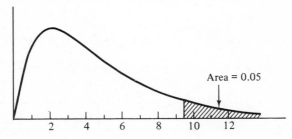

Fig. 6.1 Chi-square distribution with four degrees of freedom showing the P value when $X = 9.49$

is adequate if less than one out of five or more cells or two out of ten or more cells have expected frequencies as small as two provided that n is large so that all other e_i's are greater than 5.0. The test is two-sided even though only the upper tail of the distribution is utilized as the rejection region. Fig. 6.1 shows the P value for a given value of X calculated from Eq. (6.6) where the degrees of freedom are assumed to be four.

Example 6.4 Consider the problem raised in the beginning of this section regarding an experiment on a genetic theory. The postulated ratios of $1:2:1$ can be reformulated into the following hypothesis:

$$H_0: \quad p_1 = \tfrac{1}{4}, \qquad p_2 = \tfrac{1}{2}, \qquad p_3 = \tfrac{1}{4}.$$

The observed frequencies are $n_1 = 14$, $n_2 = 28$, and $n_3 = 18$. From the relation $e_i = np_{i0}$, the expected frequencies are: $e_1 = 15$, $e_2 = 30$, and $e_3 = 15$. Thus,

$$X = \frac{(14 - 15)^2}{15} + \frac{(28 - 30)^2}{30} + \frac{(18 - 15)^2}{15} = 0.80.$$

With two degrees of freedom, there is no reason to reject the theory at the 5% level since $0.80 < 5.99$. We can conclude that the observed frequencies are in agreement with theoretical distribution. ▲

6.5 Two-by-Two Table

Suppose a set of frequency data is observed in a two-way classification with two rows and two columns as shown in Table 6.1. Such a table is known as a *two-by-two table,* or a *fourfold table.*

TABLE 6.1 Two-by-two table

	C	D	Total
A	a	b	$a + b$
B	c	d	$c + d$
Total	$a + c$	$b + d$	$n = a + b + c + d$

A probability model for the data of Table 6.1 is shown in Table 6.2.

TABLE 6.2 Probability model for two-by-two table

	C	D	
A	p_{11}	p_{12}	$p_{1.}$
B	p_{21}	p_{22}	$p_{2.}$
	$p_{.1}$	$p_{.2}$	1

In Table 6.2, p_{11} is the probability that the two events A and C occur together, p_{12}, that A and D occur together, etc.; p_1. is the probability that A occurs, $p_{.1}$, that C occurs, etc. To be precise, using the notation of Sec. 1.3, we have

$$p_{11} = P(A \cap C), \qquad p_{12} = P(A \cap D),$$
$$p_{21} = P(B \cap C), \qquad p_{22} = P(B \cap D),$$

and

$$p_1. = P(A), \qquad p_2. = P(B), \qquad p_{.1} = P(C), \qquad p_{.2} = P(D).$$

Note that $p_1.$, $p_2.$, etc., are the unconditional probabilities, and they can be expressed as $p_1. = p_{11} + p_{12}, p_2. = p_{21} + p_{22}$, etc.

There are two broad situations in which frequency data can be put in a two-by-two table and analyzed: (a) n independent subjects are classified into a particular row or column according to some characteristics or results of an experiment. As an example, assume that 50 individuals are classified according to colors of hair and eyes as in Table 6.3.

TABLE 6.3 Classification of 50 individuals by colors of hair and
eyes

	Light hair	Dark hair	Total
Light eyes	17	5	22
Dark eyes	6	22	28
Total	23	27	50

For the data of the above form it would be of interest to test the independency of two classifications. From (1.8), if the events in the row heading and the events in the column heading are independent, we have

$$P(A \cap C) = P(A) P(C),$$
$$P(A \cap D) = P(A) P(D), \text{ etc.}$$

The hypothesis of independency, using the notation of Table 6.2, can be formulated as:

$$H_0: p_{ij} = p_i. p_{.j}, \quad i = 1, 2; \quad j = 1, 2. \tag{6.7}$$

(b) The second situation arises when two population proportions are being compared. Suppose that two machines denoted A and B are used to produce n_1 and n_2 items which contain a and c defectives, respectively. The result can be put in the form of Table 6.1 where $n_1 = a + b$ and $n_2 = c + d$. In this case, the appropriate hypothesis can be formulated as:

$$H_0: p_1 = p_2 \tag{6.8}$$

where p_1 and p_2 represent now the expected proportion of defective productions of the two machines. (See Sec. 6.3.)

Actually, there is not much difference between the two situations. After all, if we test the hypothesis that the rows and columns of Table 6.1 are independent, we are in fact testing the hypothesis that the proportions of defective and nondefective items are independent of the populations. Therefore, hypotheses (6.7) and (6.8) are equivalent. In either situation, we use the same chi-square test which is based on Theorem 6A.

For Table 6.1, the probabilities $p_{i.}$ and $p_{.j}$ can be estimated as follows:

$$\hat{p}_{1.} = \frac{a+b}{n}, \qquad \hat{p}_{2.} = \frac{c+d}{n},$$

$$\hat{p}_{.1} = \frac{a+c}{n}, \qquad \hat{p}_{.2} = \frac{b+d}{n}.$$

Since there are n observations in all, we would expect np_{ij} observations in the ith and jth classifications. When the H_0 is true, the expected frequencies from (6.7) are given by $np_{i.}p_{.j}$ and are estimated by

$$e_{ij} = n\hat{p}_{i.}\hat{p}_{.j}.$$

Therefore, for example, the expected frequency in the first row and first column is calculated as

$$e_{11} = n\hat{p}_{1.}\hat{p}_{.1} = n\frac{(a+b)}{n}\frac{(a+c)}{n} = \frac{(a+b)(a+c)}{n}.$$

Four expected frequencies corresponding to the observed frequencies of Table 6.1 are summarized in Table 6.4 where the marginal totals equal the corresponding marginal totals of Table 6.1. The chi-square test we are discussing is adequate as long as each of the four expected frequencies of Table 6.4 is not too small. Contrary to some misconceptions, the test will be valid even when $a = 0$ or $d = 0$ or both are zero, for example.

TABLE 6.4 Expected frequency corresponding to observed frequency of Table 6.1

	C	D	Total
A	$(a+b)(a+c)/n$	$(a+b)(b+d)/n$	$a+d$
B	$(c+d)(a+c)/n$	$(c+d)(b+d)/n$	$c+d$
Total	$a+c$	$b+d$	n

Substituting the observed frequencies a, b, c, and d of Table 6.1 and the corresponding expected frequencies of Table 6.4 into (6.6), we obtain after some algebraic manipulation

$$X = \frac{(ad-bc)^2 n}{(a+b)(c+d)(a+c)(b+d)} \tag{6.9}$$

which has an approximate χ^2-distribution with one degree of freedom if each of the four expected frequencies of Table 6.4 is not too small.

Remembering that the distribution of (6.9) is only approximately a chi-square unless n is very large, Yates proposed the so-called continuity correction in order to improve the approximation. The chi-square test criterion proposed by Yates is

$$X = \sum_{i=1}^{2} \sum_{j=1}^{2} \frac{(|n_{ij} - e_{ij}| - \frac{1}{2})^2}{e_{ij}},$$

where $n_{11} = a$, $n_{12} = b$, $n_{21} = c$, and $n_{22} = d$. This formula reduces to the following form in contrast to (6.9):

$$X = \frac{\left(|ad - bc| - \frac{n}{2}\right)^2 n}{(a + b)(c + d)(a + c)(b + d)} \tag{6.10}$$

The hypothesis of independence formulated in (6.7) or the equality of two proportions formulated in (6.8) is to be rejected if the value of X given by (6.10) is greater than the critical level of the χ^2-distribution with one degree of freedom. Thus, we reject the hypothesis at $P < 0.05$ if $X > 3.84$, $P < 0.01$ if $X > 6.63$, $P < 0.001$ if $X > 10.83$, etc.

It is worthwhile to note a few important points about the above χ^2 test: (a) Although, strictly, the test is valid only for large samples, the result will be satisfactory if $n > 10$ and if each expected frequency given by Table 6.4 is at least two. (b) If $n \leq 10$, the test is not very efficient, and the so-called Fisher's exact test is to be used. See Ref. 3. (c) The test is two-sided even though only the upper tail of the χ^2-distribution is referred to for a rejection region.

Finally, we observe that the value calculated using (6.10) will always be smaller than that using (6.9). This would mean that we will be less likely to reject a hypothesis if (6.10) is used instead of (6.9). In this regard, many investigators in support of (6.9) have criticized (6.10) as being overly conservative and the issue has generated considerable controversy. There is evidence that hypothesis tests using (6.10) tend to be conservative and thus result in some inefficiency; however, in this text, a test using (6.10) is recommended. It can happen that a test based on (6.9) yields an erroneously smaller P value than it ought.

The application of the χ^2 test on a two-by-two table will be illustrated next using some examples.

Example 6.5 Let us examine whether the color of eyes is related to the hair color based on the data presented in Table 6.3. Although not needed, the corresponding expected values under the hypothesis of independency are computed (in parentheses) and compared with the observed frequencies.

	Light hair	Dark hair	Total
Light eyes	17 (10.12)	5 (11.88)	22
Dark eyes	6 (12.88)	22 (15.12)	28
Total	23	27	50

There are large discrepanices between the expected and observed data which leads to a suspicion of nonindependency. From (6.10) we compute

$$X = \frac{[|(17)(22) - (5)(6)| - 25]^2 50}{(22)(28)(23)(27)} = 13.30.$$

The result is highly significant: the color of eyes is significantly related to the color of hair with $P < 0.001$. ▲

Example 6.6 In a cancer study, 40 mice with leukemia were treated with a drug (Cytoxan), and 14 mice were alive at the end of 13 months. Eighty-five leukemic mice were used as a control group (without the therapy), and 12 survived the period. Is there a significant difference in the survival rate at 13 months? The data can be arranged in a two-by-two table as follows:

	Survived	Dead	Total
Cytoxan	14	26	40
Control	12	73	85

Using (6.10) we calculate $X = 5.99$, and we can say that the two survival rates are significantly different at $P < 0.05$. ▲

6.6 Contingency Tables

In this section we shall deal with a generalization of the two-by-two table discussed in the previous section. Often the situation is such that the frequency data can be classified according to two criteria as in Table 6.5 which consists of r rows and c columns. Such a table is known as an *r-by-c contingency table*.

In Table 6.5 n_{ij} denotes the number observed to belong to the ith row and the jth column, and R_i and N_j represent the total observed numbers in ith row and in the jth column, respectively. The total sample size is denoted by n.

A frequent question raised with the contingency table is whether or not the two criteria may be considered to be independent of each other. For instance, we may like to know whether the educational level of a group of people is independent of their political philosophy. In this case we could take a sample of n independent individuals and, for example, classify the

TABLE 6.5 Contingency table with r rows and c columns

		\multicolumn{4}{c}{Columns}	Row total
		1	2 · ·	c	
Rows	1	n_{11}	n_{12} · ·	n_{1c}	R_1
	2	n_{21}	n_{22} · ·	n_{2c}	R_2
	·	·	· · ·	·	
	·	·	· · ·	·	
	·	·	· · ·	·	
	r	n_{r1}	n_{r2} · ·	n_{rc}	R_r
Column total		N_1	N_2 · ·	N_c	n

frequency data into a three-by-two table; three rows for three educational levels and two columns for conservative and liberal.

The analysis of the r-by-c contingency table is a straight forward extension of that for a two-by-two table. By p_{ij}, denote the probability that the event will fall in the ith row and jth column. Let $p_{i.}$ and $p_{.j}$ be the unconditional probability that the event will fall in the ith row and that in the jth column, respectively. The hypothesis of independency can be formulated as

$$H_0: p_{ij} = p_{i.}p_{.j}, \qquad i = 1, 2, \ldots, r; \qquad j = 1, 2, \ldots, c. \qquad (6.11)$$

As in Table 6.4, the expected frequency value under the H_0 corresponding to n_{ij} is estimated by

$$e_{ij} = n\hat{p}_{i.}\hat{p}_{.j} = n\frac{R_i}{n}\frac{N_j}{n} = \frac{R_iN_j}{n}. \qquad (6.12)$$

The test statistic based on Theorem 6A is

$$X = \sum_{i=1}^{r}\sum_{j=1}^{c}\frac{(n_{ij} - e_{ij})^2}{e_{ij}}, \qquad (6.13)$$

which has an approximate χ^2-distribution for large n if the H_0 is true. The number of degrees of freedom (d.f.) is given by

$$\text{d.f.} = (r - 1)(c - 1).$$

Note that the greater the discrepancy between the observed and expected frequencies, the larger the value of X. The decision to reject the H_0 is made if X exceeds the critical value based on a stated significance level. Again the test would be satisfactory if the minimum expected value is about 2.0 provided that n is large so that relatively few expectations are less than five (say in one cell out of five or more cells). Usually, this condition is satisfied by pooling two or more rows, or two or more columns, thus reducing the dimension of the contingency table if some expected frequencies are too small.

Example 6.7 Table 6.6 presents data on 160 female deaths due to cancer from three countries classified according to the cancer sites. Do these data indicate that the distribution of the fatal cancer sites is different among the three countries?

TABLE 6.6 Fatal cancer sites in females for three countries

Countries	Colon	Breast	Stomach	Oral	Others	Total
U.S.	11	15	3	1	40	70
Japan	5	3	22	1	29	60
England	5	7	3	0	15	30
Total	21	25	28	2	84	160

The hypothesis is that the two classifications are independent, in this case, that the distribution of the fatal cancer sites is the same for all three countries.

If we calculate the expected frequencies for oral cancer column using (6.12), we get from top to bottom

$$\frac{(70)(2)}{160} = 0.875, \qquad \frac{(60)(2)}{160} = 0.750, \qquad \frac{(30)(2)}{160} = 0.375.$$

These expected frequencies are too small to make a valid chi-square test. In order to use the test, we pool the last two columns resulting in the three-by-four contingency table presented in Table 6.7 which also shows the expected frequencies in parentheses. The expected frequencies are calculated using (6.12): for example, for the first column, we calculate

$$\frac{(70)(21)}{160} = 9.19, \qquad \frac{(60)(21)}{160} = 7.88, \qquad \frac{(30)(21)}{160} = 3.94.$$

TABLE 6.7 Fatal cancer sites in females for three countries-pooled data

Countries	Colon	Breast	Stomach	Others	Total
U.S.	11 (9.19)	15 (10.94)	3 (12.25)	41 (37.63)	70
Japan	5 (7.88)	3 (9.38)	22 (10.50)	30 (32.25)	60
England	5 (3.94)	7 (4.69)	3 (5.25)	15 (16.13)	30
Total	21	25	28	86	160

Substituting the 12 observed frequencies n_{ij} and the 12 corresponding expected frequencies e_{ij} into (6.13), we have

$$X = \frac{(11 - 9.19)^2}{9.19} + \frac{(5 - 7.88)^2}{7.88} + \cdots + \frac{(15 - 16.13)^2}{16.13}$$

$$= 29.76.$$

There are $(r - 1)(c - 1) = (2)(3) = $ six degrees of freedom. The critical value of the χ^2-distribution at $\alpha = 0.001$ with six degrees of freedom is 22.46. Hence, the evidence shows that the distribution of the fatal cancer site is significantly different among the three countries. The probability that the chi-square value of 29.76 could be obtained by chance, if the distribution of the site is the same, is less than 0.001. (The data given in Table 6.6 is artificially constructed on the basis of the estimated figures published by *World Health Statistics Annual*, 1970–1971.) ▲

Computer programs are available to perform the chi-square test on the contingency tables. See Chap. 10.

6.7 Comparison of Several Proportions

The methods of comparing two proportions have been discussed already: the one-sided test in Sec. 6.3 and the two-sided test in Sec. 6.5 where the chi-square test was used.

When more than two proportions are to be compared, a test for an overall difference in proportions is given by the same chi-square test of Sec. 6.6. However, we wish to briefly review the test applied to the present problem.

Suppose there are c independent groups of observations and that in the ith group with N_i observations, n_i has a certain characteristic, say, being "positive." The data can be summarized in a two-by-c table as shown in Table 6.8.

TABLE 6.8 Layout of data for comparing c proportions

Group	1	2	·	·	·	c	Total
Positive	n_1	n_2	·	·	·	n_c	R_1
Negative	$N_1 - n_1$	$N_2 - n_2$	·	·	·	$N_c - n_c$	R_2
Total	N_1	N_2	·	·	·	N_c	n

Let p_i be the true proportion of positives for the ith group. The null hypothesis which is desired to be tested is

$$H_0: p_1 = p_2 = \cdots p_c,$$

against the alternative that any two or more proportions are different.

The application of (6.12) gives the expected frequencies corresponding to the observed frequencies of Table 6.8, and (6.13) provides an appropriate chi-square test statistic. For a two-by-c table under consideration, a convenient computing formula can be obtained after some manipulation on (6.13) as follows:

$$X = \frac{\sum_{i=1}^{c} (n_i^2/N_i) - R_1^2/n}{PQ} \tag{6.14}$$

where $P = R_1/n$ and $Q = 1 - P$. This statistic, X, is distributed approximately as the χ^2-distribution with $c - 1$ degrees of freedom, under the H_0, when n is large.

As for the use of (6.13), the test based on the formula (6.14) is adequate if the minimum expected frequency is greater than about 2.0 provided that n is large, so that most expected frequencies are large.

It is important to realize that a rejection of the hypothesis indicates simply that the overall difference in proportions is significant, but does not show which proportions are different from which others. The following ad hoc method is suggested if the latter is the real objective. The first step is to test the overall difference using (6.14). Only if the result is significant do we proceed to apply the chi-square test for two-by-two table given by (6.10) on any combination of two columns of Table 6.8. It must be borne in mind that these tests are not independent, and for this reason the test for the overall difference is essential. For example, if we use a 5% critical value and then test a number of proportions separately, the actual significance level will be greater than 5%. The first stage test guards against falsely rejecting the hypothesis. If the overall difference is not significant, it is neither useful nor valid to perform the test on each pair of proportions separately for possible differences at the given significance level.

Example 6.8 In an experiment in cancer research, 162 mice, each with a transplanted tumor, were divided randomly into four groups, and each was treated with one of four different radiation doses. Table 6.9 summarizes the number of cured and the total number of animals in each of the four groups. (The data are extracted from an experiment performed by Dr. Carlos Perez of Washington University.) We wish to test the hypothesis that the cure rate is the same for all four groups.

Using (6.14) we calculate

$$X = \frac{10^2/42 + 32^2/41 + 37^2/39 + 32^2/40 - 111^2/162}{(111/162)(1 - 111/162)}$$
$$= 55.65.$$

The degrees of freedom are $c - 1 = 3$, for which the critical value at $P = 0.001$ point is 16.27. Since $X > 16.27$, the departure from the hypothesis of a common cure rate is thus significant at $P < 0.001$.

TABLE 6.9 Number of cured and uncured mice in four groups

| | Radiation dose in rads | | | | |
	2000	3000	4000	5000	Total
Cured	10	32	37	32	111
Not cured	32	9	2	8	51
Total	42	41	39	40	162

Next, using the ad hoc method suggested, we can show that the cure rate with 2000 rads is significantly different from each of the other three groups at $P < 0.001$, whereas the latter three groups are not significantly different from each other at the α-level of 0.05. ▲

6.8 Two-by-Two Table for Paired Observations

The application of a two-by-two table discussed in Sec. 6.5 includes the comparison of two population proportions. One basic assumption in using the test is that two sets of samples are independent. Consider the experimental situation in which the individual members of one sample are paired with particular members of the other sample. We must clearly distinguish between this situation and that where two samples are not paired and are independent. In the paired case, the two corresponding observations tend to be correlated, and hence, the assumption of independency cannot be validly made. Accordingly, a different analysis is required to compare two proportions.

In a comparative study of two populations, it could happen that some extraneous factors cause a significant difference, whereas there was no difference. Similarly, extraneous factors could also hinder us from detecting the difference if there was a difference. For example, consider a medical investigation designed to compare the cure rate of a certain disease using two different drugs. It is clearly conceivable that the response from any individual will depend on the extraneous factors such as sex, age, weight, seriousness of the disease, previous therapy, etc., besides the real effect of the drugs. Pairing is a technique frequently used in order to create homogeneity within pairs with regard to the extraneous factors.

As an illustration, suppose two treatments are compared in a clinical trial in which the result is simply improvement or no improvement. If there are 50 paired observations, then the result may look like Table 6.10 where 1

TABLE 6.10 Paired binomial observations

Observation	Treatment A	Treatment B
1	1	0
2	1	1
3	0	1
4	0	0
.	.	.
.	.	.
.	.	.
50	1	1
Total number of "1"	29	23

and 0 represent improvement and no improvement, respectively. Suppose that, as indicated in Table 6.10, the total number of 1 is 29 for A and 23 for B.

Let p_A and p_B denote the probabilities of improvement by the two treatments, and suppose we are interested in testing the hypothesis:

$$H_0: p_A = p_B.$$

In testing the H_0, we may naively present the data in the form of Table 6.11 and try to analyze it using (6.10), but this is a misapplication of the formula.

TABLE 6.11 Incorrect summary of paired observations into two-by-two table

	1	0	Total
Treatment A	29	21	50
Treatment B	23	27	50
Total	52	48	100

It is still valid to estimate the proportions p_A and p_B by $\frac{29}{50} = 0.58$ and $\frac{23}{50} = 0.46$, respectively, from Table 6.11. However, in order to compare the two proportions p_A and p_B, the data should be presented in the form of Table 6.12.

TABLE 6.12 Correct summary of paired data for comparing two proportions

		Treatment A		
		0	1	Total
Treatment B	0	19	8	27
	1	2	21	23
Total		21	29	50

It turns out that the pertinent information, as far as testing the hypothesis $H_0: p_A = p_B$ is concerned, is contained in the other diagonal of Table 6.12, that is, by 8 and 2 in the above example. For the obvious reason, the entries in the main diagonal are called *tied frequencies* and the other diagonal *untied frequencies*. The reason that the tied frequencies do not enter into the analysis is seen from the following:

$$\begin{aligned} p_A - p_B &= P(A = 1) - P(B = 1) \\ &= [P(A = 1, B = 1) + P(A = 1, B = 0)] \\ &\quad - [P(A = 1, B = 1) + P(A = 0, B = 1)] \\ &= P(A = 1, B = 0) - P(A = 0, B = 1). \end{aligned}$$

Thus, in comparing the two proportions based on the paired data, we only need to analyze the untied frequencies. For this reason the paired data can be summarized most conveniently in the form of Table 6.13.

TABLE 6.13 Information needed in comparing two proportions based on paired data

"Success" by A but "failure" by B	a
"Success" by B but "failure" by A	b
Total	$a + b$

If there is no difference between p_A and p_B, we would expect $E(a) = E(b)$. Suppose that $a > b$. For testing the hypothesis H_0 against a one-sided alternative $H_1: p_A > p_B$ the correct P value is given by the probability of observing b or less "successes" in $a + b$ trials or, equivalently, the probability of getting a or more successes in $a + b$ trials when the probability of a success is $\frac{1}{2}$. Thus,

$$P = \sum_{x=0}^{b} \binom{a + b}{x} \left(\frac{1}{2}\right)^x \left(\frac{1}{2}\right)^{a+b-x} = \left(\frac{1}{2}\right)^{a+b} \sum_{x=0}^{b} \binom{a + b}{b}. \tag{6.15}$$

If $a > b$ and the alternative is $H_1: p_A < p_B$, then the hypothesis cannot be rejected, and there is no need for any analysis. For $H_1: p_A < p_B$ if $a < b$, then we only need to replace b with a in (6.15). The exact two-sided P value for the alternative $H_1: p_A \neq p_B$ is obtained by twice the P value of (6.15).

Note that the hypothesis $H_0: p_A = p_B$ is equivalent to $H_0: p_A = \frac{1}{2}$ and $p_B = \frac{1}{2}$. Thus, if $a + b$ is large, an approximate two-sided test can be performed using a chi-square test of goodness of fit. Using (6.6), we obtain the following convenient test statistic which has a χ^2-distribution with a single degree of freedom:

$$X = \frac{(a - b)^2}{(a + b)}. \tag{6.16}$$

Example 6.9 Let's compare the two treatments A and B based on the data presented in Table 6.12. From (6.15), we compute

$$P = \left(\frac{1}{2}\right)^{10} \sum_{x=0}^{2} \binom{10}{2} = 0.055.$$

For the two-sided test the P value is 0.11, and hence, the two treatments are indicated to be not significantly different according to the evidence. ▲

The generalization of the situation to the case of more than two treatments is given by the Cochran's Q test. See Ref. 3.

*6.9 Test of Hypothesis and Sample Size

The reason and purpose of determining an appropriate sample size have already been discussed in Sec. 5.9. We shall now consider the problem of determining the sample size required in testing hypotheses, so that the probability of rejecting H_0 when it is true be limited to α, while the probability of rejecting H_0 if it is false be $1 - \beta$.

6.9.1 Sample Size for Testing a Proportion

Consider the problem of testing the following hypotheses at α level of significance:

$$H_0: p = p_0 \text{ vs. } H_1: p > p_0.$$

We would like to determine n so that the probability of rejecting H_0 is at least $1 - \beta$ when the true p is greater than or equal to $p_1 = p_0 + \delta$. Assuming a large n, using the same argument of Sec. 5.6 and the normal approximation given by (6.1), we can write

$$\alpha = P(\text{Reject } H_0 \,|\, H_0 \text{ is true})$$

$$= P\left(\frac{\hat{p} - p_0}{\sqrt{p_0 q_0/n}} > z_\alpha\right)$$

and

$$\beta = P\,(\text{Accept } H_0 \,|\, H_1 \text{ is true}) = P\left(\frac{\hat{p} - p_0}{\sqrt{p_0 q_0/n}} \leq z_\alpha \,|\, q = p_1\right)$$

$$= P\left(\frac{\hat{p} - p_1}{\sqrt{p_1 q_1/n}} \leq z_\alpha \sqrt{\frac{p_0 q_0}{p_1 q_1}} - \frac{\delta}{\sqrt{p_1 q_1/n}}\right).$$

Setting

$$z_\beta = z_\alpha \sqrt{\frac{p_0 q_0}{p_1 q_1}} - \frac{\delta}{\sqrt{p_1 q_1/n}}$$

where z_β is such that $P(Z < z_\beta) = \beta$, and solving for n, we obtain the formula

$$n = \frac{(z_\alpha \sqrt{p_0 q_0} - z_\beta \sqrt{p_1 q_1})^2}{\delta^2}. \tag{6.17}$$

For testing $H_0: p = p_0$ against $H_1: p < p_0$, an almost identical argument shows that n given by (6.17) is appropriate if H_0 is to be rejected with the probability of at least $1 - \beta$ when $p \leq p_0 - \delta$. For the two-sided alternative, if it is desired to reject H_0 with the probability of at least $1 - \beta$ when $p \leq p_0 - \delta$ or $p \geq p_0 + \delta$, the appropriate formula for n is

$$n = \frac{(z_{\alpha/2} \sqrt{p_0 q_0} - z_\beta \sqrt{p_1 q_1})^2}{\delta^2} \tag{6.18}$$

where $p_1 q_1$ denotes the maximum of $(p_0 - \delta)(1 - p_0 + \delta)$ and $(p_0 + \delta)(1 - p_0 - \delta)$.

Example 6.10 Consider the typical sampling inspection problem discussed in Example 6.1. Suppose we want to have at least a 90% guarantee of rejecting the lot if the proportion of defectives is greater than or equal to 4% instead of the acceptable figure of 2%. The α level was 0.05. Determine the appropriate sample size for inspection.

We are given $\alpha = 0.05$, $\beta = 0.10$, and $\delta = p_1 - p_0 = 0.02$. Substituting $z_\alpha = 1.645$, $z_\beta = -1.282$ together with $p_0 = 0.02$ and $p_1 = 0.04$ to (6.17), we calculate $n = 579.7$. Thus, about 580 units should be inspected for the purpose. ▲

Example 6.11 Consider the planning of a clinical trial to compare two analgesics A and B. Each subject with chronic headache receives both analgesics in a random sequence in separate time periods and is asked to indicate his preference. The investigator would like to be at least 90% sure of detecting a difference of 20% or greater while controlling the probability of a false positive finding of 5%. How many patients are needed?

Let p be the preference probability of A. The appropriate hypotheses are $H_0 : p = 0.5$ and $H_1 : p \neq 0.5$. It is specified that $\alpha = 0.05$, $\beta = 0.10$, and $\delta = 0.2$. In this example, $(p_0 - \delta)(1 - p_0 + \delta) = (p_0 + \delta)(1 - p_0 - \delta) = 0.21$, and thus, $p_1 q_1 = 0.21$. We calculate, using (6.18),

$$n = \frac{(1.96\sqrt{0.25} - (-1.282)\sqrt{0.21})^2}{(0.2)^2} = 39.1.$$

About 39 patients are required. ▲

6.9.2 Sample Size for Comparing Two Proportions

In the study designed to compare two proportions, p_1 and p_2, a frequent question deals with the required sample size. Specifically, the problem is to determine the adequate sample size which guarantees the probability of at least $1 - \beta$ for detecting the difference between the two proportions when the difference is greater than or equal to δ. We shall examine the problem for the one-sided test discussed in Sec. 6.3. The sample size problem for the two-sided test or, equivalently, the chi-square test for a two-by-two table is too complicated to be dealt with here.

In order to simplify the situation, we shall assume that equal sample size n, i.e., $n = n_1 = n_2$, using the notation of Sec. 6.3, is to be taken from each of two populations. Although the development of the approximate sample size formula is quite analogous to that for (6.17), it is somewhat more involved. Even so, such a formula including (6.17) is valid only if n is large. Fortunately, a table has been constructed using a more exact method and is extracted in Table A13. Note that the entry in Table A13 is for testing $p_1 = p_2$ against $p_1 < p_2$. It is noted that $H_0 : p_1 = p_2$ and $H_1 : p_1 < p_2$ are equivalent to $H_0 : q_1 = q_2$ and $H_1 : q_1 > q_2$, respectively, so that the table can be adapted for the case when $p_1 > 0.5$.

Example 6.12 Suppose in planning a clinical experiment as in Example 6.6, that two independent groups of patients are to be used and that the first group is to receive a standard treatment and the second a new drug. It is hoped to detect the difference with the probability 0.95 if the cure rate from the new drug is at least 25% more than that from the standard treatment. Assuming that the rate using the standard drug is at most 20%, determine the required sample size for each of the two groups. The significance level of $\alpha = 0.05$ is to be used.

Letting $p_1 = 0.20$ for the standard drug and $\delta = 0.25$, we read $n = 89$ from Table A13 when $\alpha = \beta = 0.05$. If nothing is known about the p_1, then assuming the least favorable situation when $p_1 = 0.35$ or 0.40, the sample size for each group is obtained as $n = 100$.

For the sake of further illustration, suppose that the cure rate from the standard drug is at least 0.70. This means the failure rate is at most 0.30, and that we wish to detect the difference in the failure rate of at least 25%. The required sample size is obtained as $n = 98$.   ▲

6.10 Summary

A large part of this chapter dealt with the methods of testing various hypotheses using frequency data. The single most useful tool was a statistic called chi-square, which has a χ^2-distribution. The various tests discussed in this chapter are summarized in Table 6.14.

TABLE 6.14 Summary of tests based on frequency data

Test	Section	Formula no.
About a single proportion	6.2	(6.1–6.3)
Comparison of two proportions		
Unpaired case		
One-sided	6.3	(6.5)
Two-sided	6.5	(6.10)
Paired case	6.8	(6.15)
Comparison of three or more proportions	6.7	(6.14)
Independency for two-by-two table	6.5	(6.10)
Independency for contingency table	6.6	(6.13)
Goodness of fit	6.4	(6.6)

PROBLEMS

1. According to a poll based on a sample of 760 voters, 46% approved the performance of a president. (a) Test the hypothesis that the population proportion approving the president was 0.5 against the alternative that it was below 0.5. (b) Determine the 95% confidence interval for the corresponding population percentage.

2. In one experiment on genetics, Mendel crossed two types of peas and counted the seeds of plants as shown below. A theory of genetics states that the frequencies should be in the ratios 9 : 3 : 3 : 1.

Yellow and round	315
Yellow and wrinkled	101
Green and round	108
Green and wrinkled	32

Does this data support the theory?

3. The figures on lead-paint deaths in St. Louis for 15 years were compiled by the St. Louis Public Health Laboratory as follows:

Number of deaths	Number of years
0	6
1	6
2 or more	3

Test whether the number of deaths can be described by a Poisson distribution with a <u>rate</u> of one per year. $\lambda = 1 \, yr.$

4. A study was made to examine the possible influence of including token payment on response to mail questionnaires. The table below summarizes the result:

	Responded	Not responded	Total
Payment included	46	54	100
Payment not included	132	68	200

Does this data suggest there is a significant influence of payment on responses?

5. A seed company claims that 90% of its radish seeds germinate. Out of 100 seeds planted, 14 failed to germinate. Formulate a hypothesis about the company's claim, and perform a test of the hypothesis.

6. In a recent survey on religion conducted in the U.S. and Japan, the following question was asked of 100 Americans and 50 Japanese: do you believe in life after death? Sixty-six percent of Americans and 18% of Japanese answered yes. Assuming that the result was based on random samples, test to see whether the proportions of persons who believe in life after death were significantly different between the two countries.

7. In a study conducted to determine a possible relationship between career goals and smoking habits of college students, a sample of 247 individuals was classified as follows. (1958 *Archive of Internal Medicine*, 101, p. 377.)

Career goals	Nonsmokers	Light smokers	Heavy smokers
Science	15	11	5
Nonscience	45	84	87

(a) Test whether there is a significant relation between career goals and the three categories of smoking habits. (b) Compare nonsmokers and smokers (both light and heavy smokers combined) as to the difference in their career objectives.

8. A meteorologist makes a study of the frequency of the occurrence of ice formation on aircraft, classifying the results by frontal systems involved. The data obtained is summarized as follows:

	No icing occurred	Icing occurred	Total
Cold frontal system	118	24	142
Occluded frontal system	96	44	140
Warm frontal system	222	54	276
Total	436	122	558

Set up the hypothesis to test whether the probability of the occurrence of icing is the same with all frontal systems. Test the hypothesis.

9. The following data were obtained from three experiments with each experiment based on the response of ten dogs to two brands of dog food, *A* or *B*. In the experiment each animal was placed in the center with the two foods placed equidistant from him.

The first preference

	A	*B*	Total
Exp. No. 1	4	6	10
Exp. No. 2	3	7	10
Exp. No. 3	2	8	10
	9	21	30

Compare the two dog foods based on the given data.

10. A sample of 200 fourteen-year-olds in the St. Louis public school system is selected, and the children are classified according to their weight and blood pressure. Are weight and blood pressure independent?

	Blood pressure			
	Less than 99	99–110	110–120	Over 120
Less than 102	10	20	11	5
102–132	6	42	35	26
Over 132	0	6	15	24

11. Four boxes of different brands of canned tuna containing 24 cans each were examined for quality. Many cans were found to contain filth such as parts of bugs, human hair, etc. The numbers of cans in each crate with such filth were: 14, 11, 4, 12. Can you conclude that the four brands were equally clean?

12. In a series of autopsies on 199 heavy smokers, evidence of hypertension was found in 38.2%. For 288 moderate, 152 light, and 161 nonsmokers, the corresponding percentages were 40.3, 45.5, and 50.3, respectively. Is hypertension independent of the smoking category? (From 1962 *Science*, 138, p. 975.)

13. A sample of 100 arthritis patients was used in a double-blind clinical experiment to compare two different drugs denoted *A* and *B*. The two drugs were given to each patient in a random order in two different time periods. The responses of 100 patients are summarized:

$$\begin{array}{ll} A \text{ is better} & 22 \\ B \text{ is better} & 27 \\ \text{No difference} & 51 \end{array}$$

What can you conclude from this data as to the efficacy of the two drugs?

14. In a poll conducted at several large universities, the following question was asked: do you think it is wrong for people to have sexual relations before marriage? Among 87 freshmen, 23% said yes, and 77% said no. Of 75 sophomores polled, the responses were 20% yes and 80% no. Of 140 juniors and seniors, only 13% said yes, and 87% replied no. Test whether there is a significant difference between the three groups with respect to their views on premarital sex.

15. The sex distribution of the children of 30 families with exactly three children each is given as follows:

No. of sons	0	1	2	3
No. of families	4	10	12	4

(a) Name a suitable distribution for describing the distribution of the number of sons for the three-child family.

(b) Test the rationality of the above distribution using the given data.

16. A company is considering the purchase of a new machine which makes bolts. The company will buy the machine if the manufactured bolts requiring rework is 10% or less. In a sample of 20 bolts produced by the machine, four required rework.

(a) Does the company want to purchase the machine if the significance level is fixed at 5%?

(b) Suppose the company wishes to have at least a 95% chance of not buying the machine if the real proportion of rework is 12%. Was the sample of 20 enough for the analysis?

17. It is suspected that in six-cylinder engines, the rate of deterioration of the front cylinder is different from the other five cylinders.

(a) Suppose that in 120 six-cylinder cars, 26 have the front cylinder in the worst condition. Calculate the P value of the hypothesis H_0: $p = \frac{1}{6}$ against the appropriate alternative.

(b) If the sample size is to be chosen so that the chance of rejecting the H_0 when in fact $p = \frac{1}{5}$ is 90%, what should the sample size be? Let $\alpha = 0.2$ in a two-sided test.

*18. Consider the planning of a medical experiment to compare two possible anticancer agents using mice implanted with tumor cells. Two groups of equal numbers of mice are to be used.

(a) Suppose the numbers of survivors after four weeks are to be compared. Of course, the survival rate is unknown, but it is estimated to be between 25% to 65% for both groups. It is desired to detect a difference of $d = 20\%$ with the probability of 0.90 using a test with the significance level of 0.10. What should the sample size be for each group if a one-sided test, described in Sec. 6.3, is to be used?

(b) The same question but with $d = 10\%$.

(c) In an experiment with 40 mice which received the drug A, 25 survived four weeks, while 13 out of 40 survived in the other groups treated with the drug B. Analyze the result.

19. Consider the statistics of flying-bomb hits in a part of London during World War II. The total area was divided into 576 small areas of 0.25 square kilometers each, and the number n_k of areas with exactly k hits was recorded in the following table. (From 1946 *Journal of the Institute of Actuaries*, 72, p. 48.) Test the hypothesis that the data follows a Poisson distribution with the mean $\lambda = 0.93$.

k	0	1	2	3	4 and over
n_k	229	211	93	35	8

*20. Show that the two-sided test based on the statistic Z described in Sec. 6.3(c) for testing $H_0: p_1 = p_2$ is equivalent to the chi-square test given by (6.9). (Hint: Show that the chi-square statistic is equal to the square of the Z.)

*21. Derive the formula given by (6.9) by substituting the observed and expected frequencies into (6.6).

*22. Obtain the formula given by (6.16).

7

Estimation and Testing Hypotheses:
Measurement Data

7.1 Introduction

In this and the next two chapters we shall discuss a number of estimators and tests widely used in analyzing data where for the most part observations are made on a continuous scale. A large portion of the procedures of this chapter is concerned with testing hypotheses about a single population and about differences between two populations.

The procedures described in this chapter can be classified broadly into three categories. The first and main category deals with methods of analysis of data from a normal distribution. The primary justification and reason for concentrating our attention on methods relating to the normal distribution is that this distribution is suitable, at least approximately, for a large number of practical applications. The second and more practical reason is that the methods based on the normal distribution are extensive and are well developed to facilitate and simplify the solution of many practical problems. Because the normal distribution involves two parameters, the mean and variance, we shall deal with the problems concerning these parameters. For example, a problem may be to test whether a sample can be considered to have been obtained from a population having a given mean; or, based on two independent samples, we may wish to test whether they could reasonably have been obtained in random sampling from populations with identical means.

There are situations when the normal distribution model simply is not appropriate even approximately. When very little or nothing is known about the distribution of the basic variable, and when the only assumption is that the density function is continuous, the second category of procedures is applicable. The third category which does not require the assumption of the normal distribution consists of approximate procedures applicable when the sample size is large. The justification of the third category of procedures is based on the Central Limit Theorem which states that for large sample size n, $\bar{X}$ has practically a normal distribution.

7.2 Variance of a Normal Distribution

The inference about the variance, σ^2, of a normal distribution arises in connection with the study of variability of a population.

Let X_1, X_2, ..., X_n be observations from a normal distribution $N(\mu, \sigma^2)$ with unknown variance σ^2. The unbiased estimate of σ^2 is given by the sample variance which is calculated as

$$s^2 = \frac{\sum (X_i - \bar{X})^2}{n - 1} \tag{7.1}$$

where $\bar{X}$ is the sample mean. For the estimator of σ, we shall use the sample standard deviation which is the square root of (7.1).

The basic theory, on which the inference of the variance of a normal distribution is based, is given by the following theorem.

Theorem 7A

If X is normally distributed with the population variance, σ^2, and s^2 is the sample variance based on n observations, then $(n - 1)s^2/\sigma^2$ has a χ^2-distribution with $n - 1$ degrees of freedom.

By means of this theorem, we can determine a confidence interval, or we can test a hypothesis about σ^2 based on a sample variance s^2.

7.2.1 Confidence Interval for σ^2

Consider the problem of finding the $100(1 - \alpha)\%$ confidence interval for σ^2 or for σ. From Table A3, for given degrees of freedom, we can find two values, $\chi^2_{1-\alpha/2}$ and $\chi^2_{\alpha/2}$, such that

$$P(\chi^2_{1-\alpha/2} < \chi^2 < \chi^2_{\alpha/2}) = 1 - \alpha. \tag{7.2}$$

Since $(n - 1)s^2/\sigma^2$ has a χ^2-distribution we can write

$$P\left(\chi^2_{1-\alpha/2} < \frac{(n - 1)s^2}{\sigma^2} < \chi^2_{\alpha/2}\right) = 1 - \alpha \tag{7.3}$$

from which the $100(1 - \alpha)\%$ confidence interval of σ^2, after some algebraic manipulation, is obtained as

$$\frac{(n - 1)s^2}{\chi^2_{\alpha/2}} < \sigma^2 < \frac{(n - 1)s^2}{\chi^2_{1-\alpha/2}}. \tag{7.4}$$

We are $100(1 - \alpha)\%$ confident that this interval will cover the unknown variance in the sense that the above inequality will hold true $100(1 - \alpha) \%$ of the time in the long run in repeated sampling.

The confidence interval for the unknown standard deviation σ of a normal distribution is obtained by taking the positive square roots of the two limits given by (7.4).

7.2.2 Test Concerning a Variance

Consider the problem of testing the hypothesis that the variance σ^2 of a normal distribution equals some specified value σ_0^2; that is, $H_0: \sigma^2 = \sigma_0^2$. Frequently in applications we are interested in testing the hypothesis against the one-sided alternative $\sigma^2 < \sigma_0^2$ or perhaps $\sigma^2 > \sigma_0^2$. The problem arises, for example, when we wish to test the variability of a manufacturing process or the reliability of a measuring device. For instance, a manufacturing process may be stopped for an adjustment if the variance of a measurement of the item produced exceeds a certain specified σ_0^2. The hypothesis and its alternative may be formulated as $H_0: \sigma^2 = \sigma_0^2$ and $H_1: \sigma^2 > \sigma_0^2$ in that case.

We shall consider the test under three different formulations of the alternative H_1. The test statistic is $(n-1)s^2/\sigma_0^2$ which according to Theorem 7A has a chi-square distribution with $n-1$ degrees of freedom when σ_0^2 is the true variance.

(a) $H_0: \sigma^2 = \sigma_0^2$ vs. $H_1: \sigma^2 < \sigma_0^2$.

If $\sigma^2 < \sigma_0^2$, the sample variance s^2 is likely to be smaller than σ_0^2, causing the test criterion to fall toward the lower tail of the chi-square distribution, and this would favor H_1 instead of H_0. Thus, we reject H_0 if

$$\frac{(n-1)s^2}{\sigma_0^2} < \chi_{1-\alpha}^2 \tag{7.5}$$

where $\chi_{1-\alpha}^2$ is the lower $100\alpha\%$ point of a chi-square distribution with $n-1$ degrees of freedom and α denotes the desired significance level.

(b) $H_0: \sigma^2 = \sigma_0^2$ vs. $H_1: \sigma^2 > \sigma_0^2$.

We would reject H_0 in favor of H_1 if

$$\frac{(n-1)s^2}{\sigma_0^2} > \chi_{\alpha}^2. \tag{7.6}$$

(c) $H_0: \sigma^2 = \sigma_0^2$ vs. $H_1: \sigma^2 \neq \sigma_0^2$.

This is a two-sided test. We would reject H_0 if either

$$\chi^2 = \frac{(n-1)s^2}{\sigma_0^2} < \chi_{1-\alpha/2}^2 \quad \text{or} \quad \chi^2 > \chi_{\alpha/2}^2. \tag{7.7}$$

For example, for a two-sided test with significance level of 5% when $n = 10$, the hypothesis H_0 is rejected if $(10-1)s^2/\sigma_0^2$ is either less than 2.70 or greater than 19.02.

The three different regions of rejection for testing $H_0: \sigma^2 = \sigma_0^2$, depend-

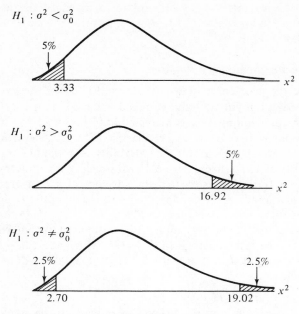

Fig. 7.1 Rejection regions for testing $H_0: \sigma^2 = \sigma_0^2$

ing on the three different forms of the H_1, are sketched in Fig. 7.1. It is assumed that $\alpha = 0.05$ and the degrees of freedom are nine.

The curves shown in Fig. 7.1 are the graphs of the chi-square distribution with nine degrees of freedom and are the exact shape of the distribution of the test statistic when the H_0 is true. Thus, when the H_0 is true and if the given rejection regions are used, the chance of falsely rejecting the hypothesis H_0 would be 5% in each case.

Example 7.1 A quality control engineer of a liquor company is concerned about the variability in the amount of liquor in 36-ounce bottles. Suppose that variability greater than 0.2, specified in terms of the standard deviation, would not be acceptable, and that a sample of 12 bottles from a population gave $s = 0.3$.

(a) Test whether the standard deviation is equal to the specified value of 0.2.

(b) Construct the 95% confidence interval for the true standard deviation.

The first problem can be formulated as

$$H_0: \sigma^2 = 0.04 \quad \text{vs.} \quad H_1: \sigma^2 > 0.04.$$

We compute $11(0.09)/0.04 = 24.75$. At the $\alpha = 0.05$ level of significance, we reject the hypothesis, H_0, since $\chi^2_{0.05} = 19.68$ at 11 degrees of freedom. Thus, variability appears to be greater than the specified magnitude.

The 95% confidence interval for σ^2 is computed to be

$$\frac{(11)(0.09)}{21.92} \quad \text{to} \quad \frac{(11)(0.09)}{3.82}$$

or 0.045 to 0.259. Taking the square roots of these numbers, the desired 95% confidence interval for σ is obtained as 0.21 to 0.51. ▲

7.3 Mean of a Normal Distribution with Unknown Variance

Let $X_1, X_2, \ldots, X_n$ be random observations from a normal distribution $N(\mu, \sigma^2)$ where both μ and σ^2 are unknown. We shall describe methods of constructing the confidence interval and of testing the hypothesis regarding the mean μ.

The important theorem which facilitates the methods is given by the following.

Theorem 7B

If $X_1, X_2, \ldots, X_n$ is a sample of size n from the normal distribution with mean μ, the variable

$$t = \frac{\bar{X} - \mu}{s/\sqrt{n}} \qquad \longleftarrow \text{——}(7.8)$$

has a t-distribution with $n - 1$ degrees of freedom.

Theorem 7B is a general result: a normal variable which does not involve the unknown variance divided by the sample standard deviation sometimes can be expressed in a form which has a t-distribution.

The variable given by (7.8) resembles the standard normal variable applied in Sec. 5.6, namely,

$$Z = \frac{\bar{X} - \mu}{\sigma/\sqrt{n}}.$$

The only difference is that t involves the sample standard deviation s while Z contains the population standard deviation, σ. Therefore, if σ is unknown, the value of Z cannot be calculated, whereas the value of t can be computed provided that the mean μ is known. This is the reason why t is so important: it enables us to solve problems dealing with the mean of the normal distribution when its standard deviation is unknown.

7.3.1 Confidence Interval for μ

In order to determine the $100(1 - \alpha)\%$ confidence interval for μ we find from Table A2 the value $t_{\alpha/2}$, for given degrees of freedom $n - 1$, so that

$$P(-t_{\alpha/2} < t < t_{\alpha/2}) = 1 - \alpha. \tag{7.9}$$

Substituting (7.8) to (7.9), we can write

$$P\left(-t_{\alpha/2} < \frac{\bar{X} - \mu}{s/\sqrt{n}} < t_{\alpha/2}\right) = 1 - \alpha, \tag{7.10}$$

which yields the desired confidence interval:

$$\bar{X} \pm t_{\alpha/2} \frac{s}{\sqrt{n}}. \tag{7.11}$$

The exact upper $100\alpha\%$ point of the t-distribution, namely, the point t_{α} which satisfies $P(t > t_{\alpha}) = \alpha$, can be read from Table A2 if the degrees of freedom is equal to or less than 30. When the degrees of freedom v is greater than 30 but less than 500, it is adequate to employ approximate interpolation in the table to determine the point t_{α}. Because of the symmetry of the t-distribution about zero, like the standard normal curve, the lower $100\alpha\%$ point is given by $-t_{\alpha}$. For v greater than or equal to 500, it is satisfactory to assume v is equal to 500 for practical purposes. Alternatively, it is satisfactory to assume that $t_{\alpha} = z_{\alpha}$, where z_{α} is the $100\alpha\%$ of the standard normal distribution if v is very large because the t-distribution converges to the standard normal distribution as v approaches infinity.

7.3.2 Test Concerning a Mean: One-Sample t-test

Consider the problem of testing the hypothesis that the mean μ of a normal distribution with unknown variance is equal to a specified value μ_0; that is, we may wish to test $H_0: \mu = \mu_0$ on the basis of a sample of size n. For example, we may wish to check the validity of the claim made on a brand of cigarettes that its mean nicotine content is 0.9 milligrams, or that an advertised mileage per gallon of a certain make of automobile is 30 miles, and so on. In most practical situations, the variance is unknown; the case when the variance is known has been described in Sec. 5.6.

The test statistic to be used is

$$t = \frac{\bar{X} - \mu_0}{s/\sqrt{n}} \tag{7.12}$$

which has a t-distribution with $n - 1$ degrees of freedom, according to Theorem 7B, when μ_0 is the true mean.

Suppose that either we are quite certain that if the mean is not equal to the postulated value μ_0 under H_0, then its value must be smaller than μ_0, or that we are not concerned if the mean is greater than μ_0. The appropriate formulation of the hypothesis takes the form:

(a) $H_0: \mu = \mu_0$ vs. $H_1: \mu < \mu_0$.

The smaller the sample mean $\bar{X}$ compared to μ_0, the less faith we should have in the truth of H_0 and the more confidence for H_1. To be precise, the rejection region of H_0 at level α is determined to be

$$t = \frac{\bar{X} - \mu_0}{s/\sqrt{n}} < -t_\alpha, \qquad (7.13)$$

where t_α is the upper $100\alpha\%$ point of the t-distribution with $n - 1$ degrees of freedom.

Next, consider the test of the hypothesis H_0 against the alternative H_1 of the following form:

(b) $H_0: \mu = \mu_0$ vs. $H_1: \mu > \mu_0$.
In this case we reject H_0 in favor of H_1 if $\bar{X}$ is large enough to conclude with a high degree of confidence that H_0 is false. To be precise, we reject H_0 if

$$t > t_\alpha. \qquad (7.14)$$

In most common cases the problem is formulated as a two-sided test.

(c) $H_0: \mu = \mu_0$ vs. $H_1: \mu \neq \mu_0$.
We wish to reject H_0 in this case if $\bar{X}$ is very much different from the postulated mean μ_0. At level α, we reject H_0 if either

$$t < -t_{\alpha/2} \quad \text{or} \quad t > t_{\alpha/2}. \qquad (7.15)$$

Each of the three tests given above is known as a *one-sample t-test*.

The rejection regions for the t-test for the above three formulations of the hypotheses are displayed in Fig. 7.2. In the figures, it is assumed that $\alpha = 0.05$ and the degrees of freedom are 19.

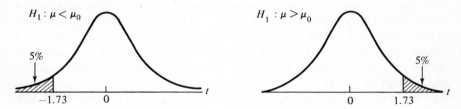

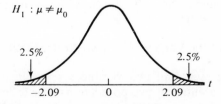

Fig. 7.2 Rejection regions for testing $H_0: \mu = \mu_0$ when σ is unknown

The graphs shown in Fig. 7.2 are the t-distribution with 19 degrees of freedom and are the exact shape of the distribution of the test statistic when the H_0 is true. Thus, if the rejection regions of the figures are used in the test, the risk of an erroneous decision to reject the H_0 when it is true is 5% in each case.

Example 7.2 A pharmaceutical company is concerned about filling its bottles with either too little or too much medicine. A filling machine, which has been set to fill 8-ounce bottles, is checked by selecting at random nine bottles and measuring their contents. The results are $\bar{X} = 8.05$ and $s = 0.035$. Assuming that the quantity filled by the machine is normally distributed, test the hypothesis that the mean μ is 8 ounces as it should be.

Since the company is concerned about either underfilling or overfilling, a two-sided test is appropriate. The hypothesis and the alternatives can be formulated as

$$H_0: \mu = 8.0,$$
$$H_1: \mu \neq 8.0.$$

Using (7.12), we calculate

$$t = \frac{8.05 - 8.0}{0.035/\sqrt{9}} = 4.29.$$

The number of degrees of freedom is $9 - 1 = 8$. At 8 degrees of freedom, the result is significant at $P < 0.01$, since $4.29 > 3.355$. The company can conclude that the mean quantity filled by the machine is significantly different from 8 ounces. Sampling variation does not provide a reasonable explanation of the difference between $\bar{X} = 8.05$ and $\mu = 8.0$. Incidentally, if the company is only concerned about overfilling, the alternative should be formulated as $H_1: \mu > 8.0$, and (7.14) provides a test. On the other hand, if it is only concerned with underfilling, then $H_1: \mu < 8.0$ is the proper alternative, and the test is given by (7.13).

The 95% confidence interval for μ can be calculated from (7.11). The upper $t_{0.025}$ point of the t-distribution with 8 degrees of freedom is 2.306. The desired confidence interval is

$$8.05 \pm (2.306)(0.035)/\sqrt{9}$$

or 8.023 to 8.077 ounces. Note that this confidence interval does not contain $\mu = 8.0$ hypothesized by the H_0. ▲

7.4 Comparison of the Variances of Two Normal Distributions

The inference regarding two variances arises in connection with the comparison of variabilities of two populations. The following theorem is the fundamental basis for the statistical analysis dealing with variances of two normal distributions.

Theorem 7C

Let s_1^2 and s_2^2 be sample variances based on independent samples of sizes n_1 and n_2, respectively, from two normal distributions with the true variances, σ_1^2 and σ_2^2. The variable

$$F = \frac{s_1^2/\sigma_1^2}{s_2^2/\sigma_2^2}$$

has the F-distribution with $(n_1 - 1, n_2 - 1)$ degrees of freedom.

It is not necessary to assume that the two distributions have equal means. The point to remember is that the above F-ratio has two numbers of degrees of freedom, $n_1 - 1$ for the numerator sample variance and $n_2 - 1$ for the denominator sample variance. The application of Theorem 7C follows next.

7.4.1 Confidence Interval for σ_1^2/σ_2^2

Suppose we have independent samples of size n_1 and n_2 from two normal distributions $N(\mu_1, \sigma_1^2)$ and $N(\mu_2, \sigma_2^2)$, respectively, and we wish to determine the $100(1 - \alpha)\%$ confidence interval for the ratio σ_1^2/σ_2^2.

From Theorem 7C we know that $(s_1^2/\sigma_1^2)/(s_2^2/\sigma_2^2)$ has an F-distribution with $(n_1 - 1, n_2 - 1)$ degrees of freedom. We wish to find two numbers, $F_{1-\alpha/2}$ and $F_{\alpha/2}$, from Table A4 such that

$$P\left(F_{1-\alpha/2} < \frac{s_1^2/\sigma_1^2}{s_2^2/\sigma_2^2} < F_{\alpha/2}\right) = 1 - \alpha. \tag{7.16}$$

Only the upper percentage points of F are found in Table A4. The lower percentage points $F_{1-\alpha/2}$ may be obtained as reciprocals of the $F_{\alpha/2}$ with $(n_2 - 1, n_1 - 1)$ degrees of freedom denoted by $F_{\alpha/2}(n_2 - 1, n_1 - 1)$. That is,

$$F_{1-\alpha/2}(n_1 - 1, n_2 - 1) = \frac{1}{F_{\alpha/2}(n_2 - 1, n_1 - 1)}. \tag{7.17}$$

From (7.16) the $100(1 - \alpha)\%$ confidence interval for σ_1^2/σ_2^2 is obtained as

$$P\left(\frac{s_1^2}{s_2^2 F_{\alpha/2}} < \frac{\sigma_1^2}{\sigma_2^2} < \frac{s_1^2}{s_2^2 F_{1-\alpha/2}}\right) = 1 - \alpha.$$

7.4.2 Test for the Equality of Two Variances

As has been previously noted, a situation occasionally arises where the variabilities of two populations are to be compared. For example, of two brands of light bulbs with equal mean lifetimes, we would prefer the brand with the smaller variance in lifetime. A social scientist may be interested in comparing the variances of the incomes in two different regions to compare the degree of heterogeneity in income distributions. We shall discuss the test of the hypothesis $H_0: \sigma_1^2 = \sigma_2^2$ on the basis of the sample variances s_1^2 and s_2^2 calculated from n_1 and n_2 observations, respectively.

The test, given under three different formulations of the alternative H_1, is based on Theorem 7C. The ratio s_1^2/s_2^2, according to the theorem, has the F-distribution with $(n_1 - 1, n_2 - 1)$ degrees of freedom if $\sigma_1^2 = \sigma_2^2$.

(a) $H_0: \sigma_1^2 = \sigma_2^2$ vs. $H_1: \sigma_1^2 < \sigma_2^2$.

A relatively small value of s_1^2/s_2^2 will occur if s_1^2 is significantly smaller than s_2^2, suggesting $\sigma_1^2 < \sigma_2^2$. To be precise, for a given probability of error α, we reject H_0 if

$$F = \frac{s_1^2}{s_2^2} < F_{1-\alpha} \qquad (7.18)$$

where $F_{1-\alpha}$ is the $100(1 - \alpha)\%$ point of F-distribution with $(n_1 - 1, n_2 - 1)$ degrees of freedom. The chance of falsely rejecting the hypothesis H_0 using the decision rule, if the H_0 is true, will be $100\alpha\%$. Note that the critical value $F_{1-\alpha}$ has to be determined from the relation (7.17).

(b) $H_0: \sigma_1^2 = \sigma_2^2$ vs. $H_1\ \sigma_1^2 > \sigma_2^2$.

At a given level α, we reject H_0 if

$$F > F_\alpha. \qquad (7.19)$$

(c) $H_0: \sigma_1^2 = \sigma_2^2$ vs. $H_1: \sigma_1^2 \neq \sigma_2^2$.

The hypothesis H_0 is likely to be false if s_1^2 and s_2^2 are significantly different or, equivalently, if the value of F is either too small or too large, since the probability of obtaining such an extreme value will be small if H_0 is true. At level α, we reject H_0 if either

$$F < F_{1-\alpha/2} \text{ or } F > F_{\alpha/2}. \qquad (7.20)$$

We can state that if s_1^2 and s_2^2 satisfy (7.20), s_1^2 and s_2^2 differ significantly at the α level of significance.

Example 7.3 (a) The systolic blood pressure of 33 healthy 15-year-old boys showed $\bar{X}_1 = 125$ and $s_1 = 9.8$. The corresponding figures for 22 15-year-old girls were $\bar{X}_2 = 115$ and $s_2 = 11.0$. Test at the 2% level the hypothesis $\sigma_1^2 = \sigma_2^2$ against $\sigma_1^2 \neq \sigma_2^2$ (b) Compare the standard deviation s_1 of the blood pressure for 15-year-old boys with $s_3 = 20.0$ for the blood pressure of adult males 45–59 years old based on the sample size of 120.

(a) According to (7.20), the critical region is

$$\frac{s_1^2}{s_2^2} < \frac{1}{2.55} = 0.39 \quad \text{or} \quad \frac{s_1^2}{s_2^2} > 2.72.$$

Since $s_1^2/s_2^2 = 0.79$, we do not reject the hypothesis of a common variance (or a standard deviation).

(b) We calculate $s_3^2/s_1^2 = 4.16$. At the 2% level the rejection region for the hypothesis $\sigma_1^2 = \sigma_3^2$ is $s_3^2/s_1^2 < 0.47$ or $s_3^2/s_1^2 > 1.86$, and therefore we reject the hypothesis of a common variance. The implication is that the dispersion of the blood pressure of the middle aged male is significantly greater than that of younger boys. In other words, the blood pressure of the older group is significantly more heterogeneous than that of the younger group. ▲

7.5 Comparison of Means of Two Normal Distributions: Independent Case

Problems involving two means arise perhaps more frequently than any other statistical problem. Let $N(\mu_1, \sigma_1^2)$ and $N(\mu_2, \sigma_2^2)$ be the distributions of two populations where μ_1, μ_2, σ_1^2, and σ_2^2 are all unknown. Let $\bar{X}_1, s_1$ and $\bar{X}_2$, s_2 be the sample mean and sample standard deviation of the two distributions based on the sample size of n_1 and n_2, respectively, from each population.

7.5.1 Pooled Standard Deviation When $\sigma_1 = \sigma_2$

When the two variances can be assumed to be the same, then the common estimate for the standard deviation based on s_1 and s_2 is calculated from the formula

$$s_p = \sqrt{\frac{(n_1 - 1)s_1^2 + (n_2 - 1)s_2^2}{n_1 + n_2 - 2}}. \qquad (7.21)$$

This quantity is a sort of average standard deviation of s_1 and s_2. (Consider what happens to s_p when the sample sizes are equal.) The quantity s_p is called the *pooled standard deviation*. Being an average of s_1 and s_2, it satisfies the following inequality:

$$\min(s_1, s_2) \leq s_p \leq \max(s_1, s_2).$$

7.5.2 Confidence Interval for $\mu_1 \bar{-} \mu_2$ When $\sigma_1 = \sigma_2$

Based on a result similar to Theorem 7B, it is known that the variable

$$t = \frac{(\bar{X}_1 - \bar{X}_2) - (\mu_1 - \mu_2)}{s_p\sqrt{\dfrac{1}{n_1} + \dfrac{1}{n_2}}} \qquad (7.22)$$

has a t-distribution with $n_1 + n_2 - 2$ degrees of freedom.

In order to find the $100(1 - \alpha)\%$ confidence interval for $\mu_1 - \mu_2$, we find the value $t_{\alpha/2}$ so that

$$P\left(-t_{\alpha/2} < \frac{(\bar{X}_1 - \bar{X}_2) - (\mu_1 - \mu_2)}{s_p\sqrt{\dfrac{1}{n_1} + \dfrac{1}{n_2}}} < t_{\alpha/2}\right) = 1 - \alpha.$$

Rearranging the terms, we find the desired confidence interval:

$$(\bar{X}_1 - \bar{X}_2) \pm t_{\alpha/2}s_p\sqrt{\frac{1}{n_1} + \frac{1}{n_2}}. \qquad (7.23)$$

It should be emphasized that the confidence interval given by (7.23) is valid only when the two populations can be assumed to have the same variance. Sec. 7.5.4 gives the appropriate formula for the confidence interval when the two variances cannot be assumed to be the same.

7.5.3 Testing the Equality of Two Means When $\sigma_1 = \sigma_2$: Two-Sample t-Test

As has been mentioned, many problems arise where we may wish to compare two populations with regard to their means. For example, a farmer may compare the yields of two varieties of corn. A taxi company may be interested in comparing the lifetime of two brands of tires. A great part of medical research is concerned with comparing two different types of drugs or treatments in terms of means of a response variable.

Let $\bar{X}_1$ and $\bar{X}_2$ be the sample means calculated from two independent samples, one from each population. The mathematical basis for testing the hypothesis $H_0: \mu_1 - \mu_2 = d_0$, where d_0 is some specified value, is given by the t-statistic of (7.22): if $\mu_1 - \mu_2 = d_0$, the variable

$$t = \frac{(\bar{X}_1 - \bar{X}_2) - d_0}{s_P\sqrt{\dfrac{1}{n_1} + \dfrac{1}{n_2}}}$$

has a t-distribution with $n_1 + n_2 - 2$ degrees of freedom.

In most practical applications, usually, but not necessarily always, $d_0 = 0$. For example, in comparing the mean blood pressure of teenagers in two different school districts, we may be concerned only when the difference is greater than 5 mm. In this case the hypotheses can be formulated as

$$H_0: \mu_1 - \mu_2 = 5 \quad \text{against} \quad H_1: \mu_1 - \mu_2 < -5 \quad \text{or} \quad \mu_1 - \mu_2 > 5.$$

However, for simplicity we shall assume henceforth $d_0 = 0$, and we shall consider testing the equality $H_0: \mu_1 = \mu_2$ with three different formulations of the alternatives. If $d_0 \neq 0$, the reader can easily modify the given test.

First, consider the problem of testing the hypothesis against the alternative of the following form:

(a) $H_0: \mu_1 = \mu_2$ vs. $H_1: \mu_1 < \mu_2$.
It would seem reasonable to reject this H_0 in favor of H_1 if $\bar{X}_1 - \bar{X}_2$ is too small. The precise rejection region at level α is

$$t = \frac{\bar{X}_1 - \bar{X}_2}{s_P\sqrt{\dfrac{1}{n_1} + \dfrac{1}{n_2}}} < -t_\alpha \tag{7.24}$$

where t_α is the upper $100\alpha\%$ point of the t-distribution with $n_1 + n_1 - 2$ degrees of freedom.

(b) $H_0: \mu_1 = \mu_2$ vs. $H_1: \mu_1 > \mu_2$.
The hypothesis H_0 is rejected if

$$t > t_\alpha. \tag{7.25}$$

(c) $H_0: \mu_1 = \mu_2$ vs. $H_1: \mu_1 \neq \mu_2$.

Clearly, H_0 should be rejected if $\bar{X}_1 - \bar{X}_2$ is either too small or too large since this would favor H_1 instead of H_0. The rejection region for H_0 is two-sided:

$$t < -t_{\alpha/2} \quad \text{or} \quad t > t_{\alpha/2}. \tag{7.26}$$

We may state that $\bar{X}_1$ and $\bar{X}_2$ are significantly different at the α level of significance if (7.26) is satisfied. It should be emphasized that the tests given by (7.24–7.26) are not valid when $\sigma_1 \neq \sigma_2$. If the two variances are not the same, then the actual probability of Type I error could be greater than or smaller than the desired level of α. It is known that the two-sample t-test is to a large degree insensitive to the assumption of normality so the test can be applied to nonnormal populations without materially affecting the stated significance level if n_1 and n_2 are not too small. Unfortunately, on the other hand, the test is somewhat sensitive to the assumption of equal variance. In addition to the assumption of equal variances, the tests require that $\bar{X}_1$ and $\bar{X}_2$ are the sample means based on two independent sets of samples.

Example 7.4 Ten test plots of equal size planted with one variety of corn yielded on the average 6.98 bushels per plot with a standard deviation of 0.71 while eight equal sized test plots planted with a second variety of corn yielded on the average 7.56 bushels with standard deviation of 0.69. Compare the mean yields of the two types of corn.

The null and alternative hypothesis can be formulated as

$$H_0: \mu_1 = \mu_2 \quad \text{and} \quad H_1: \mu_1 \neq \mu_2,$$

where μ_1 and μ_2 denote the mean yields per plot for the two varieties of corn. Substituting the given values $s_1 = 0.71$ and $s_2 = 0.69$ into (7.21), we calculate the pooled standard deviation:

$$s_p = \sqrt{\frac{9(0.71)^2 + 7(0.69)^2}{10 + 8 - 2}} = 0.70.$$

Next, we substitute $\bar{X}_1 = 6.98$, $\bar{X}_2 = 7.56$, and $s_p = 0.70$ into (7.24), getting

$$t = \frac{6.98 - 7.56}{0.70\sqrt{\frac{1}{10} + \frac{1}{8}}} = -1.75.$$

The two-sided critical value, at the 5% level, for the t-distribution with 16 degrees of freedom is ± 2.12. Therefore, we cannot reject the hypothesis H_0; we conclude that there is no significant evidence that the two mean yields are different. In other words, the observed difference can be explained by chance or sampling variation. ▲

In the above example, it was reasonable to assume that $\sigma_1 = \sigma_2$. When the assumption is unreasonable, alternative methods must be used in testing the difference between means. The method is discussed in Sec. 7.5.5. Note that even if two populations have the identical variance, their sample variances will almost never coincide exactly because of chance variations. We can test the equality of two variances by Eq. (7.20) if their sample variances are different enough to suspect the assumption.

Example 7.5 The systolic blood pressure of 90 healthy five-year-old boys showed $\bar{X}_1 = 101$ and $s_1 = 9.6$, while the corresponding values for 33 fifteen-year-old boys were $\bar{X}_2 = 125$ and $s_2 = 9.8$. (a) Compare the mean blood pressures μ_1 and μ_2 of the two populations. Can we say that it is higher in the older boys? (b) Find the 95% confidence interval for the difference in the means.

Clearly, we can assume that the two populations have a common variance. The pooled standard deviation is, using (7.21),

$$s_p = \sqrt{\frac{(89)(9.6)^2 + (32)(9.8)^2}{90 + 33 - 2}} = 9.65.$$

The appropriate hypothesis and the alternative can be formulated as H_0: $\mu_1 = \mu_2$ and H_1: $\mu_1 < \mu_2$. Using (7.24) we calculate

$$\frac{101 - 125}{9.65\sqrt{\frac{1}{90} + \frac{1}{33}}} = -12.22,$$

which is significant with $P < 0.001$. Thus, we can state that the mean blood pressure is significantly higher in the older boys.

The 95% confidence interval for $\mu_1 - \mu_2$ is, using (7.23),

$$(101 - 125) \pm (1.98)(9.65)\sqrt{\tfrac{1}{90} + \tfrac{1}{33}}$$

or -27.89 to -20.11. ▲

7.5.4 Confidence Interval for $\mu_1 - \mu_2$ When $\sigma_1 \neq \sigma_2$

When the two population variances are not equal, the exact method for obtaining a confidence interval does not exist because neither (7.22) nor any other similar form has a t-distribution. However, it is known that the following variable

$$t = \frac{(\bar{X}_1 - \bar{X}_2) - (\mu_1 - \mu_2)}{\sqrt{v}} \tag{7.27}$$

where

$$v = \frac{s_1^2}{n_1} + \frac{s_2^2}{n_2} \tag{7.28}$$

has an approximate t-distribution with the degrees of freedom f estimated by
note V is squared

$$f = \frac{v^2}{\dfrac{(s_1^2/n_1)^2}{n_1 - 1} + \dfrac{(s_2^2/n_2)^2}{n_2 - 1}}. \tag{7.29}$$

The value of f will not be an integer, and the closest value in the table may be used instead of interpolation, since f itself is an approximation. It may be useful to note that

$$\min(n_1 - 1, n_2 - 1) \leq f \leq n_1 + n_2 - 2.$$

Using the fact that t given by (7.27) has an approximate t-distribution, the $100(1 - \alpha)\%$ confidence interval for $\mu_1 - \mu_2$ can be obtained as

$$(\bar{X}_1 - \bar{X}_2) \pm t_{\alpha/2}\sqrt{v} \tag{7.30}$$

where v is given by (7.28) and $t_{\alpha/2}$ is the $100\alpha/2\%$ point of the t-distribution with f degrees of freedom.

7.5.5 Testing the Equality of Two Means When $\sigma_1 \neq \sigma_2$: Aspin-Welch Test

As in the case of the confidence interval, there exists only an approximate method for testing $\mu_1 = \mu_2$ if the two population variances are not equal. The method is based on the fact that if $\mu_1 = \mu_2$, the variable

$$t = \frac{\bar{X}_1 - \bar{X}_2}{\sqrt{v}} \tag{7.31}$$

where

$$v = \frac{s_1^2}{n_1} + \frac{s_2^2}{n_2}$$

has an approximate t-distribution with the degrees of freedom given by (7.29).

(a) $H_0: \mu_1 = \mu_2$ vs. $H_1: \mu_1 < \mu_2$.
We reject H_0 if

$$\frac{\bar{X}_1 - \bar{X}_2}{\sqrt{v}} < -t_\alpha \tag{7.32}$$

where t_α is the $100\alpha\%$ point of the t-distribution with the degrees of freedom f defined in (7.29).

(b) $H_0: \mu_1 = \mu_2$ vs. $H_1: \mu_1 > \mu_2$.
The hypothesis H_0 is rejected if

$$\frac{\bar{X}_1 - \bar{X}_2}{\sqrt{v}} > t_\alpha. \tag{7.33}$$

(c) $H_0: \mu_1 = \mu_2$ vs. $H_1: \mu_1 \neq \mu_2$.
This is a two-sided test. We reject H_0 if either

$$t = \frac{\bar{X}_1 - \bar{X}_2}{\sqrt{v}} < -t_{\alpha/2} \quad \text{or} \quad t > t_{\alpha/2}. \tag{7.34}$$

The above three tests (7.32–7.34) are often called *Aspin-Welch tests*.

Example 7.6 The tensile strength, in pounds, of two types of cable wires A and B was being compared. Denoting the mean tensile strengths by μ_A and μ_B, the problem was formulated by the hypothesis $H_0: \mu_A = \mu_B$ against $H_1: \mu_A \neq \mu_B$. The following data were obtained: $n_A = 30$, $\bar{X}_A = 132.6$, $s_A = 5.2$ for the wire A, and $n_B = 25$, $\bar{X}_B = 129.3$, $s_B = 2.7$ for wire B. Carry out the test assuming that the population variances are different.

The appropriate test is given by (7.34). We calculate

$$v = \frac{(5.2)^2}{30} + \frac{(2.7)^2}{25} = 1.19$$

and

$$t = \frac{132.6 - 129.3}{\sqrt{1.19}} = 3.03.$$

The approximate number of degrees of freedom computed using (7.29) is

$$f = \frac{(1.19)^2}{\dfrac{[(5.2)^2/30]^2}{29} + \dfrac{[(2.7)^2/25]^2}{24}} = 44.9.$$

The computed value of $t = 3.03$ at 45 degrees of freedom gives a two-sided $P < 0.01$. Thus, we can conclude that the mean strength of cable wire A is significantly greater than that of B. ▲

7.6 Comparison of Means of Two Normal Distributions: Paired Case

In some experiments dealing with two populations, it is convenient and advantageous to take the observations in pairs. The basic purpose for pairing is to remove the possible extraneous factors that may cause a difference in means.

For example, a clinical trial of two treatments using two groups of patients may yield a significant difference due to some extraneous differences between the two groups, whereas there may be no actual difference in the treatment effects. The extraneous factors, for example, might be age, sex, weight, or severity of disease, etc. The paired experiment may be carried out by giving the two treatments to the same patient at different times or on different sites (like two eyes) depending on the nature of the clinical trial.

As another example, suppose we are interested in comparing gas mileage at two different speeds. Clearly, it is advantageous to use one group of cars driven at different speeds instead of using two groups of automobiles for the two different speeds.

Let $(X_{11}, X_{12}), (X_{21}, X_{22}), (X_{31}, X_{32}), \ldots, (X_{n1}, X_{n2})$ be the n pairs of observations. Since observations are paired, it cannot be assumed that the first and second members of each pair are independent. For example, a car with a good mileage at one speed tends to have a good mileage at a different speed. At any rate the methods of Sec. 7.5 are not appropriate for analyzing paired data.

In order to correctly analyze the data we work with differences, d_i, of each pair; that is

$$d_1 = X_{11} - X_{12}, d_2 = X_{21} - X_{22}, \ldots, d_n = X_{n1} - X_{n2}.$$

Then, the method of determining the confidence interval for $\mu_1 - \mu_2$ and that of testing the hypothesis $\mu_1 = \mu_2$ becomes identical to those described in Sec. 7.3, with $\bar{d}$ replacing $\bar{X}$ and using s_d, the standard deviation of d's, in place of s.

7.6.1 Confidence Interval for $\mu_1 - \mu_2$ in Paired Data

Let n represent the number of pairs. The $100(1 - \alpha)\%$ confidence interval for $\mu_1 - \mu_2$ is given by

$$\bar{d} \pm t_{\alpha/2} \frac{s_d}{\sqrt{n}},$$

where $\bar{d}$ and s_d are the sample mean and standard deviation calculated from $d_1, d_2, \ldots, d_n$, and $t_{\alpha/2}$ denotes the $100\alpha/2\%$ point of the t-distribution with $n - 1$ degrees of freedom.

7.6.2 Testing the Equality of Two Means: Paired t-Test

(a) $H_0: \mu_1 = \mu_2$ vs. $H_1: \mu_1 < \mu_2$.
   We reject H_0 if

$$\frac{\bar{d}}{s_d/\sqrt{n}} < -t_\alpha$$

   where t_α is the $100\alpha\%$ point of the t-distribution with $n - 1$ degrees of freedom.

(b) $H_0: \mu_1 = \mu_2$ vs. $H_1: \mu_1 > \mu_2$.
   We reject H_0 if

$$\frac{\bar{d}}{s_d/\sqrt{n}} > t_\alpha.$$

(c) $H_0: \mu_1 = \mu_2$ vs. $H_1: \mu_1 \neq \mu_2$.
   In this case, we decide to reject H_0 if either

$$t = \frac{\bar{d}}{s_d/\sqrt{n}} < -t_{\alpha/2} \quad \text{or} \quad t > t_{\alpha/2}.$$

Unlike the two-sample t-test we need not assume the two variances σ_1^2 and σ_2^2 to be equal, and this certainly is one advantage of the paired experiment. Nevertheless, the main and primary goal of pairing observations should be to remove the possible effect of extraneous variables and thereby to permit a more precise comparison. If there are no extraneous effects, the paired t-test would still be valid but less powerful compared with the two-sample t-test.

This can be seen from the fact that the degrees of freedom for the paired t-test based on n pairs of observations is $n - 1$, while it is $2n - 2$ for the corresponding unpaired test. What this means is that the paired t-test requires more observations than the "unpaired" t-test to achieve the same power.

Sometimes pairing is created artificially by the investigator who matches subjects or items on important characteristics so that members of a pair are as alike as possible with regard to these characteristics. The natural situation which yields paired data automatically is a "before-and-after" type of comparison. For example, if we study a new diet program using a sample of n individuals, the weights before and after the program yield n paired observations.

Example 7.7 Suppose that systolic blood pressure was taken from 12 hypertensive patients before and after a certain treatment. Table 7.1 (columns 1, 2, and 3) presents the results. Test whether the reduction in the blood pressure is significant.

TABLE 7.1 Blood pressure before and after treatment

Patient	Before	After	d_i
1	172	174	−2
2	166	152	14
3	189	153	36
4	164	153	11
5	158	151	7
6	197	193	4
7	182	183	−1
8	164	145	19
9	179	182	−3
10	197	197	0
11	175	159	16
12	165	151	14

Let μ_1 and μ_2 denote the mean blood pressures taken before and after the treatment. Since we are only interested in knowing whether blood pressure is reduced by the treatment, the hypothesis and the alternative can be formulated as

$$H_0: \mu_1 = \mu_2 \quad \text{and} \quad H_1: \mu_1 > \mu_2.$$

In order to analyze the data, we calculate d_i for each pair as shown in column 4 of Table 7.1. From these, we compute $\bar{d} = 9.58$ and $s_d = 11.29$, and

$$t = \frac{9.58}{11.29/\sqrt{12}} = 2.94.$$

For a one-sided test with 11 degrees of freedom, the result is significant at $P < 0.01$. We conclude that the treatment is effective in lowering the blood pressure of hypertensive patients. ▲

7.7 Robustness, Transformation, and Nonparametric Tests

For most of this chapter, we have been concerned with tests of hypothesis dealing with measured observations. One common assumption required for all the tests described so far is that the observations are made from a normally distributed population. What could happen to the result of the test if this normality assumption is violated? It turns out that even fairly large departures from the normal distribution may not seriously affect the test, at least in the case of large sample sizes. This is particularly true for all *t*-tests. This rather fortunate insensitivity of the test to the required assumption is called *robustness* of the test. This advantage is due to the Central Limit Theorem which implies that the sample mean or sum of sample observations has an approximately normal distribution for large sample sizes.

Even if the test is robust with respect to the normality assumption, it would be a good idea to see that the assumption is valid, at least approximately, especially when the sample size is not large. If the data in hand is such that the assumption of normality does not hold, there are two possible alternatives: transformation of observed values before applying the test, and nonparametric tests.

7.7.1 Transformation

Sometimes a particular transformation of the variable X will have an approximately normal distribution when X itself does not. For example, if the distribution of X looks like the figure on the left in Fig. 7.3, the distribution of $Y = \log X$ (base e) will be similar to the one on the right of Fig. 7.3. For some reasons an amazing proportion of the data encountered in the biomedical, socioeconomic, and engineering fields has a skewed distribution like the one in Fig. 7.3, so that the logarithmic transformation is often found to be satisfactory for "normalization."

Such a simple normalizing transformation does not always exist for a

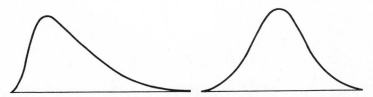

Fig. 7.3 Shapes of a distribution before and after the logarithmic transformation

nonnormal distribution. If it does, then of course the data can be analyzed using the methods described so far in this chapter after individual observations are transformed.

Example 7.8 In a cancer research study, the level of immunoglobulin, which is the protein of animal origin having antibody activity, was investigated. In the study, IgG level (in mg/ml of blood) was obtained from a sample of 20 normal children aged three to ten and from a sample of 20 children with short-term lymphoblastic leukemia. One problem was to compare the mean IgG level of the two groups. (See 1970 *Cancer*, *26*, p. 890.)

IgG levels of the 20 normal children were as follows:

13.0	14.7	11.8	8.5	9.5	10.0	11.8
13.2	14.7	11.2	17.4	12.4	18.4	10.6
16.4	11.8	11.2	14.7	10.6	13.2	

The histogram shown in Fig. 7.4 suggests that the underlying distribution is similar to the skewed curve of Fig. 7.3. The logarithmic transformations (base e) of the above 20 values are:

2.56	2.69	2.47	2.14	2.25	2.30	2.47
2.58	2.69	2.42	2.86	2.52	2.91	2.36
2.80	2.47	2.42	2.69	2.36	2.58	

The histogram using the transformed data is presented in Fig. 7.5. The figure indicates that the assumption of a normal distribution for the transformed values is valid at least approximately. The sample mean and standard deviation calculated from the transformed data are: $\bar{X} = 2.53$ and $s_x = 0.203$.

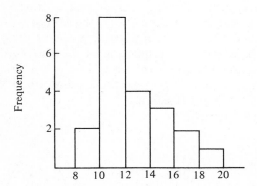

Fig. 7.4 Histogram of IgG level

Although individual values of IgG for 20 leukemia patients are not presented here, the situation was very much similar to the case for the normal. That is, the transformed values can be assumed to come from a normal dis-

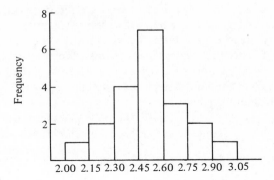

Fig. 7.5 Histogram of log-IgG (base e) level

tribution, at least approximately. The sample mean and standard deviation are calculated from the transformed data as: $\bar{Y} = 2.05$ and $s_Y = 0.344$.

First, we can test whether the two distributions have the same variance. Using the method described in Sec. 7.4.2, we obtain

$$s_Y^2 / s_X^2 = 2.87.$$

The upper 2.5% point of F-distribution with $(19, 19)$ degrees of freedom is 2.52, and hence the difference in variances is significant at two-sided $P < 0.05$. Thus, a comparison of the mean IgG level should be analyzed using the Aspin-Welch test described in Sec. 7.5.5. We calculate from (7.31)

$$t = \frac{2.53 - 2.05}{0.089} = 5.39$$

with the degrees of freedom equal to approximately 31. The difference is significant at $P < 0.001$ using the two-sided test. We can conclude that the mean IgG level (geometric mean, to be precise) for the short-term leukemia patients is significantly lower than the corresponding mean IgG level of normal children. ▲

7.7.2 Nonparametric Tests

Statistical tests which do not require the assumption of a distributional form, such as a normal distribution, are called *nonparametric*, or *distribution free*, *tests*. The t-tests are not nonparametric but parametric tests. The main advantage for considering nonparametric tests is that they can be applied without assuming the distribution form of populations.

Thus, first, although we know that the t-tests are robust with respect to the distribution assumption, the nonparametric test better controls the probability of falsely rejecting the hypothesis when the population is not normal.

Second, the rank tests we shall discuss are almost as efficient as the t-test, even when distributions are normal. If the normality assumption is correct, the t-tests are the best tests. To be specific, the rank test requires approximately 5% to 10% more observations than the t-test to be equally powerful in the case of normal distributions. However, the motivating and appealing fact is that the rank test can be much more powerful than the t-test for some nonnormal distributions.

As in the use of the t-test, the case where the two samples are independent must be distinguished from the case where the samples are paired.

7.8 Rank Test for Comparing Two Populations: Independent Case

The first nonparametric test we are going to describe is known as the *Wilcoxon two-sample rank sum test* after Wilcoxon who developed it. It is a test of the hypothesis that two distributions are identical, against the alternative that their medians are different. If $f_1(x)$ and $f_2(x)$ are the two density functions, the hypothesis H_0 is

$$H_0: f_1(x) = f_2(x).$$

In practical applications, this hypothesis may be conceived as the equality of the two medians.

Let $X_1, X_2, \ldots, X_{n_1}$ and $Y_1, Y_2, \ldots, Y_{n_2}$ denote samples of sizes n_1 and n_2 independently taken from the populations with the density functions $f_1(x)$ and $f_2(x)$, respectively. Without any difficulty, it can be assumed that $n_1 \leq n_2$. The two-sample rank test is carried out in the following steps:

(1) Arrange the combined observations from the smallest to the largest.
(2) Rank the ordered observations: give the rank of one to the smallest, the rank of two to the next smallest, etc., and the mean of the tied ranks if two or more ranks are tied.
(3) Determine the sum of ranks R for the group with the smaller sample size. Intuitively, either too small or too large a value of R suggests a contradiction of the hypothesis H_0.
(4) Reject the hypothesis H_0 if the rank sum R is too small or too large based on the critical values given in Table A6.

Table A6, where n_1 denotes the number of observations in the smaller group, can be used for n_1 and n_2 up to 20. Before describing the method for larger values of n_1 and n_2, we shall illustrate the above four steps using a simple example.

Suppose that only nine observations were made, five from Group A and and four from Group B, as follows:

$$
\begin{array}{lccccc}
A: & 17 & 8 & 19 & 13 & 10 \\
B: & 8 & 5 & -2 & 11 &
\end{array}
$$

The required steps are:

Order:	-2	5	8	8	10	11	13	17	19
Rank:	1	2	3.5	3.5	5	6	7	8	9

Rank sum of B: $R = 1 + 2 + 3.5 + 6 = 12.5$

For $n_1 = 4$ and $n_2 = 5$, one would reject H_0 at the 5% level if $R \le 11$ or $R \ge 29$ under two-sided alternative. Since our R falls between the critical points, we accept H_0 that the two populations have identical distributions.

When testing H_0 against the alternative that the first population with the smaller sample size is situated to the left of the second population, only the left-side critical region should be used (i.e., $R \le 11$ in the above example). Similarly, one should use the right-side critical region for testing against the shift being to the right. In either case, when testing against a one-sided alternative, the value of P given in Table A6 should be halved.

For larger sample sizes the distribution of R can be approximated satisfactorily by a normal distribution with the mean and variance given by the formulas

$$E(R) = \frac{n_1(n_1 + n_2 + 1)}{2},$$

$$\sigma_R^2 = \frac{n_1 n_2 (n_1 + n_2 + 1)}{12}.$$

Thus, the approximate P value can be obtained from Table A1 using the test statistic Z which has the approximate standard normal distribution:

$$Z = \frac{R - E(R)}{\sigma_R}. \tag{7.35}$$

Example 7.9 Two wires, A and B, are compared for their resistance per unit length. Sixteen tests on wire A and ten tests on B were conducted, with the following results (in ohms $\times 10^2$):

Wire A

54.7   51.4   52.5   51.9   51.3   55.3   49.6   49.6   51.7   52.1
57.9   51.8   54.9   51.6   52.8   52.5

Wire B

50.7   47.9   49.4   48.4   49.3   49.6   50.7   48.3   51.2   48.2

It is desired to test whether the resistance level is the same for the two wires. Since the sample size from wire B is smaller than that from wire A, let n_1 be the sample size for wire B: $n_1 = 10$, $n_2 = 16$. The rank sum of the observations for wire B is calculated to be

$$R = 1 + 2 + 3 + 4 + 5 + 6 + 8 + 11.5 + 11.5 + 13 = 65,$$

noting each of the three tied values of 49.6 is given the mean rank of eight, and the two tied observations of 50.7 each take the rank of 11.5. Referring to

Table A6 of the Appendix, for $n_1 = 10$ and $n_2 = 16$, the critical regions for the two-sided test are $R \leq 86$ or $R \geq 184$ at the $P = 0.01$ level. Since the calculated R value is less than 86, we reject the hypothesis of a common distribution of the two populations, and conclude that wire B has significantly lower resistance than wire A at the $P < 0.01$ level.

The present example is also used to illustrate the application of the approximate test procedure based on (7.35). We calculate

$$E(R) = \frac{10(10 + 16 + 1)}{2} = 135.0,$$

$$\sigma_R^2 = \frac{(10)(16)(10 + 16 + 1)}{12} = 360.0.$$

Substituting these values into (7.35) yields

$$Z = \frac{65.0 - 135.0}{\sqrt{360.0}} = -3.69.$$

It follows from Table A1 that $P(Z < -3.69) = 0.0001$, and thus the two-sided P value is about 0.0002. In agreement with the conclusion reached above, the resistance level of the two wires is significantly different. ▲

The rank sum test is known to be an excellent test for the types of problems being described here. The normal approximation procedure based on (7.35) is satisfactory even for moderate sample sizes, say $n_1 \geq 10$ and $n_2 \geq 10$.

7.9 Rank Test for Comparing Two Populations: Paired Case

A two-sample nonparametric test to compare two populations on the basis of a paired sample is *Wilcoxon matched-pairs signed rank test*. Let $(X_1, Y_1), (X_2, Y_2), (X_3, Y_3), \ldots, (X_n, Y_n)$ be the n pairs of observations where the $X_1, X_2, \ldots, X_n$ are from a population with the density function $f_1(x)$ and the $Y_1, Y_2, \ldots, Y_n$ are from a population with $f_2(y)$. The hypothesis to be tested is that the two populations have identical distributions:

$$H_0: f_1(x) = f_2(x)$$

and the alternative is that their medians are different. In practical applications, as in Sec. 7.8, the hypothesis may be considered as the equality of the two medians.

The signed rank test for the paired observations is carried out in the following steps:

(1) Determine the differences $d_1 = X_1 - Y_1$, $d_2 = X_2 - Y_2, \ldots, d_n = X_n - Y_n$.

(2) Arrange the absolute differences $|d_i|$ from the smallest to the largest ignoring sign of d_i.

(3) Rank the $|d_i|$: give the rank of one to the smallest, the rank of two to the next smallest, etc., and the mean of the tied ranks if two or more ranks are tied.

(4) Affix to each rank the sign of d_i, and then determine S, the smaller sum of the like-signed ranks. Intuitively, one would expect the sums of the two like-signed ranks to be about equal when the H_0 is true.

(5) Reject the hypothesis H_0 if the value of S is too small based on the critical values given in Table A5.

The critical values of S for two-sided tests are given in Table A5 for $n = 5$ to $n = 50$. The table can be adapted for use with one-sided tests by halving the value of P given.

In order to illustrate the procedure consider the following six pairs of observations:

$$A: \quad 7 \quad 5 \quad 5 \quad 6 \quad 10 \quad 8$$
$$B: \quad 3 \quad 7 \quad 3 \quad 3 \quad 6 \quad 7$$

The required steps are:

Determine d_i: 4   −2   2   3   4   1

| Rank $|d_i|$: | 1 | 2.5 | 2.5 | 4 | 5.5 | 5.5 |
| Affix sign: | 1 | −2.5 | 2.5 | 4 | 5.5 | 5.5 |

Smaller sum of the like-signed ranks: $S = 2.5$

For $n = 6$ the critical region at the $P = 0.05$ level is given by $S \leq 1$ under the two-sided alternative. Since our S is greater than 1, we accept H_0 that the two populations have identical distributions.

For larger sample sizes the distribution of S can be approximated satisfactorily by a normal distribution with the mean and variance given by

$$E(S) = \frac{n(n + 1)}{4},$$

$$\sigma_S^2 = \frac{n(n + 1)(2n + 1)}{24}.$$

Thus, the approximate P value can be obtained from Table A1 using the test statistic Z where

$$Z = \frac{S - E(S)}{\sigma_S}. \tag{7.36}$$

Example 7.10 Consider Example 7.7 dealing with the blood pressure of 12 patients. It is desired to test whether the reduction in blood pressure is significantly different from zero using the signed rank test.

First, we order all d_i's without regard to sign and rank them as follows:

$$d_t: \quad 0 \quad -1 \quad -2 \quad -3 \quad 4 \quad 7 \quad 11 \quad 14 \quad 14 \quad 16 \quad 19 \quad 36$$
$$\text{Rank:} \quad 1 \quad 2 \quad 3 \quad 4 \quad 5 \quad 6 \quad 7 \quad 8.5 \quad 8.5 \quad 10 \quad 11 \quad 12$$

The next step is to compare the sum of ranks with negative signs and that with positive signs. In this case the smaller sum of the like-signed ranks is given by $S = 2 + 3 + 4 = 9$. From Table A5 the two-sided critical value of S when $n = 12$ is 10 at $P = 0.02$ level. Since $9 < 10$, the result is significant at $P < 0.01$ testing against the one-sided alternative that the blood pressure is reduced by the treatment. We recall that the same conclusion was reached by the application of the paired t-test.

In order to illustrate the method of determining a P value on the basis of the normal approximation, we calculate

$$E(S) = \frac{12(12 + 1)}{4} = 39,$$

$$\sigma_S^2 = \frac{12(12 + 1)(24 + 1)}{24} = 162.5.$$

Therefore, we have, using (7.36),

$$Z = \frac{9 - 39}{\sqrt{162.5}} = -2.353.$$

Using the one-sided alternative, the result is significant at $P < 0.01$ since $-2.353 < -2.326$. The justification of the normal approximation is demonstrated in this instance. ▲

7.10 Other Nonparametric Tests

There are many other nonparametric tests besides the two rank tests we have described: see Ref. 5. In this section we shall discuss two other non-parametric tests which are adaptations of the methods of Chap. 6. The first test (known as the *median test*) is appropriate for analyzing two independent samples, and the second test (called the *sign test*) is for data consisting of matched pairs. The main advantage of the two tests is their simplicity. The disadvantage is that they tend to be less efficient than the rank tests described in Sec. 7.8 and 7.9.

7.10.1 Median Test

Let $f_1(x)$ and $f_2(x)$ be the two continuous density functions, and let X_1, $X_2, \ldots, X_{n_1}$ and $Y_1, Y_2, \ldots, Y_{n_2}$ denote random samples of sizes n_1 and n_2 independently taken from the two populations. Note that the situation is identical to that of Sec. 7.8. We wish to test the hypothesis that the two populations have the same distribution, namely,

$$H_0 : f_1(x) = f_2(x).$$

The median test is carried out in the following steps:

(1) Combine the two samples together and determine the sample median.
(2) Categorize each observation as lying either below or above the median, and summarize the frequencies as in the following two-by-two table. It is assumed that $n_1 + n_2$ is even, so the median is not one of the observations. If $n_1 + n_2$ is odd, then the median will be one of the observations in the sample. In this case, that observation is deleted from all consideration, excluding it also from the sample total.

	Below median	Above median	Total
Sample 1	a	b	n_1
Sample 2	c	d	n_2

(3) Use the chi-square test given by (6.10) of Sec. 6.5 for the two-sided test.

We should remember that, as in Sec. 6.5, the test described is appropriate only when n_1 and n_2 are at least about 5 each. The median test is very simple but less efficient than the corresponding rank test of Sec. 7.8.

Example 7.11 Consider Example 7.9 dealing with the problem of comparing two wires in regard to their resistance. From the given data, the sample median of 26 observations is determined to be 51.35. We can summarize the observations as lying either below or above the median as follows:

	Below median	Above median	Total
Wire A	3	13	16
Wire B	10	0	10

Using (6.10) we calculate the chi-square value of 11.22. Thus, we can conclude that the resistance level of the two wires is significantly different at $P < 0.001$ since $11.22 > 10.83$. ▲

7.10.2 Sign Test

Let $(X_1, Y_1), (X_2, Y_2), \ldots, (X_n, Y_n)$ denote n paired samples from two populations. On the basis of these observations, we want to test the hypothesis that the two populations have the same distribution:

$$H_0: f_1(x) = f_2(x).$$

The underlying situation and the hypothesis are identical to those of Sec. 7.9. The sign test is carried out in the following steps:

(1) Determine the number of pairs in which X is greater than Y and the number of pairs in which Y is greater than X. The tied observations

are deleted. The result can be summarized as in the form of Table 6.13:

Category	Number
X is greater than Y	a
Y is greater than X	b

(2) Use exactly the same test procedure described in Sec. 6.8. See (6.15) and accompanying discussion for determining the P value.

The rationale of the sign test is that if there is no difference between the two distributions, it is expected that $E(a) = E(b)$. Therefore, the technique consists of testing the probability that X greater than Y is equal to $\frac{1}{2}$. The sign test is extremely simple to apply. However, in general, it is less powerful than the signed rank test described in Sec. 7.9.

Example 7.12 Consider Example 7.7 dealing with the problem of comparing systolic blood pressure of 12 individuals before and after the treatment. Let X be the blood pressure before the treatment and Y after the treatment. Eleven untied observations can be classified as:

$$X \text{ is greater than } Y: \quad a = 8$$
$$Y \text{ is greater than } X: \quad b = 3$$

Against a one-sided alternative that the treatment is effective in reducing the blood pressure, the P value is given by

$$P = (\tfrac{1}{2})^{11} \sum_{x=0}^{3} \binom{11}{x} = 0.113.$$

Therefore, the sign test indicates that the reduction of systolic blood pressure is not significant. A comparison of this result with that obtained by using the signed rank test shows that the signed rank test is more efficient in determining the real difference. ▲

7.11 Large Sample Test for the Mean

Given any population, for all practical applications, the mean $\bar{X}$ of a random sample can be assumed to have an approximate normal distribution with the mean μ and the variance σ^2/n if n is large. In other words, the variable

$$Z = \frac{\bar{X} - \mu}{\sigma/\sqrt{n}}$$

is approximately a standard normal variable.

When the population variance is unknown and the sample size is large, say at least about 40, the value of Z would not be appreciably different if the sample standard deviation s is used in place of σ. Thus, if the sample size is large, an approximate test for the hypothesis $H_0: \mu = \mu_0$ can be made using the following test statistic with the critical value based on Table A1:

$$Z = \frac{\bar{X} - \mu_0}{s/\sqrt{n}} \qquad (7.37)$$

irrespective of the distribution.

In the same spirit, an approximate two-sample test for the hypothesis $H_0: \mu_1 = \mu_2$, if the sample sizes n_1 and n_2 are large, can be performed using the approximate standard normal variable

$$Z = \frac{\bar{X}_1 - \bar{X}_2}{\sqrt{\dfrac{s_1^2}{n_1} + \dfrac{s_2^2}{n_2}}} \qquad (7.38)$$

regardless of the distribution of the two populations. In this approximate large sample test, there is no need for assuming the two variances to be equal. In most practical applications, the test would be adequate if both n_1 and n_2 are at least about 30 each.

Because the approximate tests described here are applicable regardless of the underlying distribution, they may be considered as nonparametric tests.

The Central Limit Theorem provides the rationale for the approximate large sample tests. Just for the sake of argument, however, suppose that a population has a normal distribution. Then, the variable given by (7.37), which is identical in form to (7.12), has a t-distribution, and the exact test is described in Sec. 7.3.2. However, as readers can verify from Tables A1 and A2, the difference between the standard normal and the t-distribution, when n is large, is minimal. Thus, the test based on (7.37) would be approximately adequate when a population has a normal distribution. A similar argument holds true for the test based on (7.38).

Example 7.13 Cholesterol levels (in mg/100 ml) measured in two age groups of healthy males can be summarized with the following statistics. (From 1973 *Journal of Clinical Investigation*, Vol. 52, p. 1533.)

Age	n	Mean	Standard deviation
20–29	43	192	33
40–49	116	226	43

Use the large sample method to compare the mean cholesterol levels of the two groups.

The hypothesis to be tested is that the mean cholesterol levels of the two groups are not different. Using (7.38), we calculate

$$Z = \frac{192 - 226}{\sqrt{\dfrac{33^2}{43} + \dfrac{43^2}{116}}} = -5.29.$$

From Table A1, the two-sided P value is found to be less than 0.001. The hypothesis is rejected, and we conclude that the mean cholesterol concentration levels are significantly different with the older-age male's being higher. (It is known that in both sexes cholesterol levels increase with age up to about 60 years of age.) ▲

*7.12 Test of Hypothesis and Sample Size

As discussed previously in Sec. 5.9, determining the right sample size is a practical as well as a critical problem. Too small a sample may fail to reject a false null hypothesis, while too large a sample is wasteful of time and resources. Unfortunately, for many practical tests discussed in this chapter, there do not exist simple formulas for the required sample size.

In this section, we shall consider the problem of the sample size required in one-sample and two-sample tests of means of the normal distribution with known variance. Since the true variances are rarely known in practice such solutions may seem to have little value. However, when some a priori knowledge or information on the variance is available, the solution not only sheds some light on the problem but also provides a good idea on the required sample size in the one-sample and two-sample t-tests.

To be precise, we would like to determine the sample size n required in testing hypothesis H_0 with the probability α of rejecting H_0 if it is true, and the probability $1 - \beta$ of rejecting H_0 when it is false.

7.12.1 Sample Size for One-Sample Test of Mean

Consider the problem of testing the following hypotheses:

$$H_0: \mu = \mu_0 \quad \text{vs.} \quad H_1: \mu > \mu_0.$$

We would like to determine n so that the probability of rejecting H_0 is at least $1 - \beta$ when $\mu \geq \mu_0 + \delta$ where $\delta > 0$.

Recall from (5.6) that the test is performed by rejecting H_0 for a given sample size n, if

$$\frac{\bar{X} - \mu_0}{\sigma/\sqrt{n}} > z_\alpha$$

where z_α is the $100\alpha\%$ point of the standard normal distribution, namely, the value satisfying

$$\alpha = P(Z > z_\alpha).$$

Now, we have from (5.7), when $\mu = \mu_0 + \delta$,

$$\beta = P\left(Z \leq z_\alpha - \frac{\delta}{\sigma/\sqrt{n}}\right).$$

For given β, let z_β denote the value such that $P(Z \leq z_\beta) = \beta$. Then, we have

$$z_\alpha - \frac{\delta}{\sigma/\sqrt{n}} = z_\beta,$$

or solving for n we obtain the formula for n,

$$n = \frac{\sigma^2(z_\alpha - z_\beta)^2}{\delta^2}. \tag{7.39}$$

For testing $H_0: \mu = \mu_0$ against $H_1: \mu < \mu_0$, an almost identical argument shows that (7.39) is the appropriate formula for n if H_0 is to be rejected with the probability of at least $1 - \beta$ when $\mu \leq \mu_0 - \delta$.

Consider now the two-sided alternative $H_1: \mu \neq \mu_0$. It is desired that H_0 be rejected with the probability of at least $1 - \beta$ when $\mu \leq \mu_0 + \delta$ or $\mu \geq \mu_0 + \delta$. The test is given by rejecting H_0 if either

$$\frac{\bar{X} - \mu_0}{\sigma/\sqrt{n}} < -z_{\alpha/2} \text{ or } > z_{\alpha/2}.$$

The required sample size n is given by the formula

$$n = \frac{\sigma^2(z_{\alpha/2} - z_\beta)^2}{\delta^2}. \tag{7.40}$$

7.12.2 Sample Size for Two-Sample Test of Means

In order to simplify the problem we shall assume that the equal sample size n is to be taken from each of two populations with known variances σ_1^2 and σ_2^2, respectively. The hypothesis of interest is

$$H_0: \mu_1 = \mu_2.$$

For given sample means $\bar{X}_1$ and $\bar{X}_2$, the test would be based on the variable

$$Z = \frac{\bar{X}_1 - \bar{X}_2}{\sqrt{\dfrac{\sigma_1^2 + \sigma_2^2}{n}}}$$

which has the standard normal distribution when the H_0 is true.

The problem, for a given significance level α, is to determine n which will assure the probability of at least $1 - \beta$ of detecting the difference when the difference between the two means is greater than or equal to δ. The methods here are slight modifications of those in Sec. 7.12.1. Hence, we shall only summarize the result in Table 7.2 where z_β denotes the value such that $P(Z \leq z_\beta) = \beta$.

Example 7.14 An experiment using sick mice is planned to compare the mean survival time under two different treatments. From the previous investigation the standard deviation of both survival times is estimated to be about 1.5 (weeks). Assume that the underlying distributions are normal. How many

TABLE 7.2 Critical region and sample size required for testing H_0: $\mu_1 = \mu_2$ with different alternatives when variances are known

Alternative H_1	Critical region	Sample size n^*
$\mu_1 > \mu_2$	$\dfrac{\bar{X}_1 - \bar{X}_2}{\sqrt{(\sigma_1^2 + \sigma_2^2)/n}} > z_\alpha$	$\dfrac{(\sigma_1^2 + \sigma_2^2)(z_\alpha - z_\beta)^2}{\delta^2}$
$\mu_1 < \mu_2$	$\dfrac{\bar{X}_1 - \bar{X}_2}{\sqrt{(\sigma_1^2 + \sigma_2^2)/n}} < -z_\alpha$	$\dfrac{(\sigma_1^2 + \sigma_2^2)(z_\alpha - z_\beta)^2}{\delta^2}$
$\mu_1 \neq \mu_2$	$\dfrac{\bar{X}_1 - \bar{X}_2}{\sqrt{(\sigma_1^2 + \sigma_2^2)/n}} < -z_{\alpha/2}$ or $> z_{\alpha/2}$	$\dfrac{(\sigma_1^2 + \sigma_2^2)(z_{\alpha/2} - z_\beta)^2}{\delta^2}$

*Sample size required in each population.

mice does the experiment require if we want to detect a mean difference of one week with the probability of not less than 0.9 while guarding against the false positive with probability 0.05?

We are given $\alpha = 0.05$, $\beta = 0.1$, and $\sigma_1 = \sigma_2 = 1.5$, and thus, $z_{\alpha/2} = 1.96$ and $z_\beta = -1.28$. Substituting these values along with $\delta = 1$ into the last formula of Table 7.2 yields

$$n = \frac{(2.25 + 2.25)(1.96 + 1.28)^2}{1} = 47.2.$$

Thus, a total of about 94 mice is required. ▲

7.13 Summary

A large part of this chapter was devoted to the methods of testing various hypotheses. Table 7.3 summarizes the various tests most commonly used in applications and discussed in this chapter. It would be natural to extend the two-sample t-test of means discussed in this chapter to the comparison of three or more means. Techniques are available for handling these comparisons, and they are dealt with in Chap. 9.

PROBLEMS

1. The yield point (in units of 100 psi) for a sample of 16 steel castings from a large lot is recorded as follows:

70	68	65	67	67	65	69	70
65	62	68	72	71	70	70	65

Assuming the yield points of the population to be normally distributed (a) test the hypothesis H_0: $\sigma = 2$ at the 5% level of significance, (b) find the 90% confidence interval for σ.

TABLE 7.3 Summary of commonly used tests for measurement data

Main category	Test	Hypothesis	Test criterion	Situation
Tests for normal distributions	One-sample test of variance	$\sigma^2 = \sigma_0^2$	$\chi^2 = \dfrac{(n-1)s^2}{\sigma_0^2}$	Single sample
	One-sample t-test of mean	$\mu = \mu_0$	$t = \dfrac{\bar{X} - \mu_0}{s/\sqrt{n}}$	Single sample
	Two-sample test of variances	$\sigma_1^2 = \sigma_2^2$	$F = \dfrac{s_1^2}{s_2^2}$	Two independent samples
	Two-sample t-test of means	$\mu_1 = \mu_2$	$t = \dfrac{\bar{X}_1 - \bar{X}_2}{s_p\sqrt{(1/n_1) + (1/n_2)}}$	Two independent samples and common variance
	Two-sample test of means (Aspin-Welch test)	$\mu_1 = \mu_2$	$t = \dfrac{\bar{X}_1 - \bar{X}_2}{\sqrt{s_1^2/n_1 + s_2^2/n_2}}$	Two independent samples and unequal variances
	Paired t-test of two means	$\mu_1 = \mu_2$	$t = \dfrac{\bar{d}}{s_d/\sqrt{n}}$	Matched or paired sample
Nonparametric tests	Two-sample rank sum test	$f_1(x) = f_2(x)$	See Sec. 7.8	Two independent samples
	Matched-pairs signed rank test	$f_1(x) = f_2(x)$	See Sec. 7.9	Matched or paired sample
	Median test	$f_1(x) = f_2(x)$	See Sec. 7.10.1	Two independent samples
	Sign test	$f_1(x) = f_2(x)$	See Sec. 7.10.2	Matched or paired sample
Large sample test of means	One-sample test of mean	$\mu = \mu_0$	$Z = \dfrac{\bar{X} - \mu_0}{s/\sqrt{n}}$	Large sample, approximate test
	Two-sample test of means	$\mu_1 = \mu_2$	$Z = \dfrac{\bar{X}_1 - \bar{X}_2}{\sqrt{s_1^2/n_1 + s_2^2/n_2}}$	Two independent large samples, approximate test

2. Referring to the data of Prob. 1: (a) test the hypothesis H_0: $\mu = 70$ at the 5% level of significance, (b) find the 95% confidence interval for μ.

3. A sample of 25 employees selected at random from the employees of a large company yields a mean disabled time of 42.7 hours per year, with a standard deviation of 7.3 hours. Construct a 95% confidence interval for the average disabled time per employee.

4. From a large sample of whites and blacks in St. Louis, the systolic blood pressures of the ninth graders are obtained and are summarized by the local heart association as follows:

$$\text{White:} \quad n_1 = 970 \quad \bar{X}_1 = 111.7 \quad s_1 = 12.0$$
$$\text{Black:} \quad n_2 = 690 \quad \bar{X}_2 = 109.6 \quad s_2 = 10.6$$

   (a) Test the hypothesis H_0: $\sigma_1 = \sigma_2$.
   (b) Test the hypothesis H_2: $\mu_1 = \mu_2$.
   (c) Obtain the 99% confidence interval for the difference between the blood pressures of the two races.

5. The viscosity of two different brands of car oil, A and B, are measured and the results are:

   Brand A   10.41   10.23   10.24   10.31   10.35   10.36   10.39   10.44
   Brand B   10.56   10.45   10.39   10.50   10.68   10.48   10.51   10.63
             10.49

   (a) Compare the mean viscosity of the two brands assuming that the populations have normal distributions with equal variances.
   (b) Compare the viscosites of the two brands using a rank test.

6. Cardiac output (in ml/100 min) was measured in ten dogs before and during infusion of norepinephrine. Test the hypothesis of no change in cardiac output using (a) the t-test, and (b) a rank test.

Dog no.	Before	During
1	14.8	15.7
2	18.4	19.7
3	15.1	21.8
4	17.8	16.9
5	14.3	18.7
6	16.7	16.8
7	12.9	14.4
8	19.9	27.3
9	16.6	19.5
10	17.4	18.8

7. Brand R of automobile tires claimed to yield superior gas mileage. Twenty independent observations obtained from a simulated condition

gave the following results on the mileage per gallon using a common brand and brand R:

Brand R: 15.2   14.7   14.9   15.1   14.4   15.0   14.5   14.6
                 14.8   14.9

Common brand: 14.3   14.6   15.0   14.0   13.7   14.1   14.0   14.2
                 13.8   14.4

Can the claim of the manufacturer of brand R be supported?

8. In cancer research, an experiment using leukemic mice was performed to compare two administration schedules of the drug Cytoxan. The survival times in months for the two groups are recorded:

120 mg/kg once weekly:
8.0   8.75   7.75   7.25   8.5   6.0   8.5   6.0   13.5
6.5   10.0   9.75   10.0

60 mg/kg twice weekly:
13.0   10.75   7.75   11.25   13.75   12.0   11.5   14.25   13.75
11.0   8.75   10.5   8.75   9.25

Perform any analysis that you think is pertinent.

9. The effect of exercise on intraocular pressure (in mm Hg) was investigated using 12 individuals. Draw a pertinent conclusion after making an appropriate analysis on the data below:

Individual	Before exercise	After exercise	Individual	Before exercise	After exercise
a	26	17	g	21	15
b	10	9	h	15	11
c	14	8	i	17	11
d	12	9	j	12	13
e	15	11	k	24	15
f	12	8	l	19	14

10. The table below gives the lives in hundreds of hours of samples of electronic tubes of two brands A and B. Survival time data are usually skewed with a longer tail to the right, and a lognormal distribution is often used to model the data.

(a) Compare two brands assuming that the survival time has a lognormal distribution.

(b) Compare the brands using a rank test.

A: 32   34   35   37   42   47   59   62   78   84
B: 39   48   54   65   70   76   87   90   111   118

11. Suppose you are an FDA examiner and you would approve a certain drug if a certain chemical content of the drug is less than or equal to

1 mg per pill. A lab took a sample of nine and found that $\bar{X} = 1.2$ mg and $s^2 = 0.04$.

(a) Would you approve the drug (at the 5% level)?

(b) Was an α error made and what is the consequence of the error in the decision made in (a)?

12. Nine pregnant women were given an injection to induce labor. Their systolic blood pressures before and after the injection were:

	Blood pressure	
Patient no.	Before	After
1	134	140
2	122	130
3	136	132
4	120	126
5	130	140
6	132	138
7	140	140
8	118	122
9	142	144

Perform a test to determine whether the injection of this drug changes blood pressure.

13. Weight gains were recorded for two groups of experimental animals on high and low protein diets. Group A received a high protein diet. Group B received a low protein diet.

(A) High Protein	(B) Low Protein
0.94	0.49
0.79	0.82
0.96	0.73
0.98	0.86
1.02	0.81
1.02	0.97
1.08	1.06
0.91	0.70
1.20	0.61
1.05	0.82
$\bar{X}_A = 0.995$	$\bar{X}_B = 0.787$
$s_A^2 = 0.0119$	$s_B^2 = 0.0274$

Test the hypothesis: mean weight gain on the high protein diet is equal to mean weight gain on the low protein diet.

14. Total nitrogen content of the blood plasma of normal albino rats was measured at 37 and 180 days of age. (One group was sacrificed at the end of 37 days and one group at the end of 180 days.) The results are expressed as grams per 100 cc of plasma. At age 37 days, nine rats had 0.98, 0.83, 0.99, 0.86, 0.90, 0.81, 0.94, 0.92, 0.87 grams of nitrogen per

100 cc of blood. At age 180 days, eight rats had 1.20, 1.18, 1.33, 1.21, 1.20, 1.07, 1.13, 1.12 grams per 100 cc of blood. Test the null hypothesis: the difference in mean concentration of nitrogen at 37 and 180 days is zero.

15. The following data on the height and weight of 17- to 18-year-old girls is from a large sample survey conducted in the St. Louis metropolitan area. The sample size was 591 for the whites and 282 for the blacks.

| | (Height-in.) | | (Weight-lb.) | |
	White	Black	White	Black
Mean	64.2	63.8	124.9	127.1
S.D.	2.5	2.6	20.8	24.3

Perform any analysis that is useful for comparing the two races with regard to the height and weight.

16. According to Mendelian inheritance theory, certain crosses of peas should yield yellow and green peas in the $3:1$ ratio. In an experiment, 179 yellow and 49 green peas are obtained. Use a large sample method to test whether the deviation from the theory is significant.

17. Work Prob. 12 by means of the sign test.

18. Use the median test to work Prob. 13.

19. Sardine tags (which are internal) are frequently recovered primarily from canneries. Two canneries which processed almost identical amounts of fish recovered 74 and 95 tags, respectively. Is this difference significant? Assume that the tag recoveries follow a Poisson distribution. (Hint: Make use of the fact that a Poisson random variable X has the mean λ and variance λ, and that for a large sample size, X is approximately normal.)

20. According to a recently conducted poll, 47% of American households had a firearm in the house, compared with 41% in 1968 at the peak of the urban riot period. Assuming that each figure is based on 1600 randomly selected independent households, is it possible to state that the percentage increase from 1968 is significant? Use a large sample method.

*21. In comparing two different fibers with regard to percentage of shrinkage at a temperature of 120°C, a two-sample test is to be used with $\alpha = 0.05$. The variances in the shrinkage is considered to be the same for the two fibers. How large a sample from each of the fibers is needed if it is desired to have a 90% chance of rejecting the null hypothesis of no difference when the real difference is 0.5, and if the variance is known to be greater than 0.1 but less than 0.3?

*22. Derive the last formula for the sample size n given in Table 7.2.

8

Regression and Correlation Analysis

8.1 Introduction

Consider the situation where two or more measurements are made on each subject of a sample. The first problem we shall deal with in this chapter is that of estimating or predicting the value of a dependent variable Y on the basis of one or more fixed variables X. A dependent variable is understood to be a random variable, while the fixed variables are preassigned or predetermined. For example, we may wish to estimate the yield of a crop, Y, for different amounts of fertilizer, X. More generally, we may wish to predict a crop yield as a function of the amount of fertilizer, X_1, rainfall, X_2, temperature, X_3, and possibly other variables. Problems of this type are known as *regression problems*, and their main purpose is to estimate or predict one variable using other variables. The first step in regression analysis is to decide on the functional model of the relationship between variables. Consider the three different scatter diagrams of Fig. 8.1.

The scatter diagrams could suggest a pattern of a functional form or *regression model*, namely, the equation of the line or curve that expresses Y

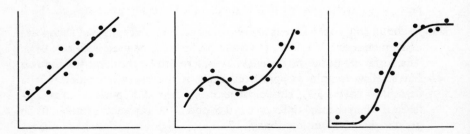

Fig. 8.1   Scatter diagrams suggesting different regression models

in terms of X. Of course, infinitely many different models are conceivable.

In this chapter we shall be concerned mainly with the most common situations where the scatter diagrams or other considerations indicate that the relationship may be linear or approximately linear.

The second problem we will be discussing is the measure of the relationship or correlation between two or more random variables rather than predicting one variable on the basis of other variables. Problems of this type are known as *correlation problems.* For example, we may be concerned with the problem of measuring and analyzing the possible relationship between the weights of father and son, or that between the IQ score and a physical characteristic of individuals, or the possible correlation between grades in a statistics course and a computer science course. As in the case of the regression model, the correlation problem can be extremely complicated depending on the functional form of the relationship. In this chapter we shall be concerned with several simple and useful measures of correlation.

8.2 Simple Linear Regression

Suppose we have n paired observations $(X_1, Y_1), (X_2, Y_2), \ldots, (X_n, Y_n)$. The simplest regression model is given by a straight line relation between Y and a single fixed variable X. To be precise, the regression model in this case is given by

$$Y = \alpha + \beta X + \epsilon \qquad (8.1)$$

where α and β are the *regression coefficients* and ϵ is an independent random variable with zero mean and the same variance σ^2 for all Y. If we denote the mean of Y for given X by $\mu_{Y \cdot X}$, then the above model can be expressed as

$$\mu_{Y \cdot X} = \alpha + \beta X.$$

In (8.1) the variable ϵ, often referred to as an error term, may be conceived as a random variable representing the fluctuation of Y for given X. Since X is a fixed variable, so is $\alpha + \beta X$, and the variables Y and ϵ have the same variance. Suppose that ϵ is normally distributed with zero mean and variance σ^2. Then, this implies that Y is normally distributed with mean $\alpha + \beta X$ and variance σ^2. The graphical representation of a simple linear regression model under the normality assumption is depicted in Fig. 8.2. Note that the mean $\mu_{Y \cdot X}$ for any given X lies on a straight line with the slope β and the intercept α.

Of course, the random variable ϵ may not have a normal distribution. We shall consider two cases depending on ϵ.

Case A: ϵ is normally distributed.

Case B: ϵ is not normally distributed.

For Case A, we shall discuss the methods of point and interval estimation and hypothesis testing, while for Case B, we shall be concerned only with

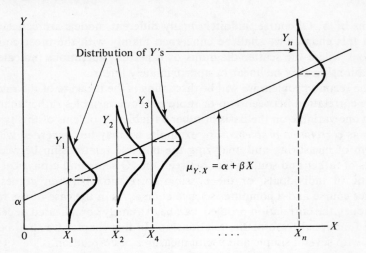

Fig. 8.2 Simple linear regression model when Y is normally distributed

point estimation because of the complexity of the problem. However, all methods are approximately adequate for Case B also if the sample size is large.

Consider the following ten pairs of observations with X representing the entrance test score and Y the statistics test score of ten students.

$$X: \quad 85 \quad 93 \quad 67 \quad 85 \quad 59 \quad 78 \quad 90 \quad 60 \quad 73 \quad 92$$
$$Y: \quad 86 \quad 93 \quad 74 \quad 80 \quad 69 \quad 85 \quad 88 \quad 64 \quad 77 \quad 86$$

A plot or scatter diagram of the above data appears in Fig. 8.3. The plot suggests that a straight line is a reasonable model in expressing Y in terms of X.

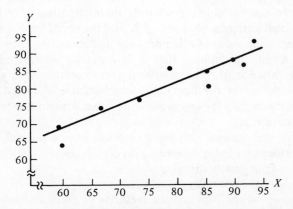

Fig. 8.3 Scatter diagram which suggests linear regression

The first problem is how best to fit a line to a set of data points. The problem is essentially that of calculating the regression line of Y on X denoted by $\hat{Y} = a + bX$ where a and b are estimators of α and β, respectively, obtained in some efficient manner.

We shall state without proof the most important theorem in regression analysis. It is the *Gauss-Markov theorem*.

Theorem 8A

For the regression model $Y = \alpha + \beta X + \epsilon$ where ϵ's for different Y's are independent with mean 0 and variance σ^2, the best estimators of α and β (in the minimum variance unbiased sense) are given by the least square estimators.

Thus, the Gauss-Markov theorem tells us how to obtain the best estimators of α and β for both Case A and Case B. For n observations, (X_1, Y_1), $(X_2, Y_2), \ldots, (X_n, Y_n)$ the least square estimators a and b for α and β are given by minimizing the quantity L given by

$$L = \sum_{i=1}^{n} [Y_i - (a + bX_i)]^2$$

with respect to a and b. This is accomplished by the usual differentiation procedure. To be precise, we set

$$\frac{\partial L}{\partial a} = -2 \sum_{i=1}^{n} (Y_i - a - bX_i) = 0,$$

$$\frac{\partial L}{\partial b} = -2 \sum_{i=1}^{n} (Y_i - a - bX_i)X_i = 0.$$

Solving the above two equations for a and b, the least square estimators of α and β are

$$a = \bar{Y} - b\bar{X}, \tag{8.2}$$

$$b = \frac{\sum (X_i - \bar{X})(Y_i - \bar{Y})}{\sum (X_i - \bar{X})^2} = \frac{\sum X_i Y_i - n\bar{X}\bar{Y}}{\sum (X_i - \bar{X})^2}. \tag{8.3}$$

Based on the above estimators the predicted $\hat{Y}$ for given X is

$$\hat{Y} = a + bX.$$

For the data of Fig. 8.3 the least square estimates using (8.2) and (8.3) are calculated to be $a = 28.13$ and $b = 0.67$. The resulting regression line $\hat{Y} = 28.13 + 0.67X$ is shown in Fig. 8.3.

In addition to estimating α and β, it is of interest to estimate σ^2. This estimator is required for the confidence interval and testing hypothesis, but it is interesting in its own right since σ^2 is the variance of Y about the true regression line $\alpha + \beta X$. An unbiased estimator of σ^2 based on the least square estimators of α and β is given by the sample variance of Y_i about the predicted value, $\hat{Y}_i$, that is,

$$s_{Y\cdot X}^2 = \frac{1}{n-2} \sum (Y_i - \hat{Y}_i)^2$$

$$= \frac{1}{n-2} \sum (Y_i - a - bX_i)^2. \tag{8.4}$$

If s_X^2 and s_Y^2 denote the sample variances of X_i and Y_i, then (8.4) can also be computed from the algebraically equivalent formula

$$s_{X\cdot Y}^2 = \frac{n-1}{n-2}(s_Y^2 - b^2 s_X^2). \tag{8.5}$$

Dividing by $n - 2$ instead of $n - 1$ makes it unbiased. Clearly, the minimum number of observations required in the regression analysis is three. The quantity $s_{Y\cdot X}$ is called the *standard error of regression line*.

Example 8.1 The age (years), systolic, and diastolic blood pressures (mm Hg) were measured for ten randomly selected patients in a children's hospital. The data are given in the following table:

Age	Systolic B.P.	Diastolic B.P.
14	100	60
1	83	51
9	112	64
7	152	65
9	104	62
12	90	60
1	92	59
3	85	51
9	120	65
1	103	62

The first problem was to determine the regression line of systolic pressure (Y) on age (X). Using (8.3) and (8.2) we have

$$a = 96.80, \qquad b = 1.11.$$

The desired regression line is

$$\hat{Y} = 96.80 + 1.11X.$$

The variances are computed to be $s_X^2 = 23.16$ and $s_Y^2 = 420.32$. Thus, using (8.5) the standard error of regression line is

$$s_{Y\cdot X} = \sqrt{\tfrac{9}{8}[420.32 - (1.11)^2(23.16)]} = 20.99.$$

It suggests fairly large variance of the systolic pressures from the regression line. ▲

8.3 Inferences about Regression Coefficients

In this section we assume that ϵ or equivalently Y are normally distributed. The methods described in this section, strictly speaking, apply only to Case A. It is known, however, that the methods are approximately valid for Case B also, provided the sample size is large.

The inferences about α and β are based on the fact that the functions

$$t = \frac{a - \alpha}{s_a} \tag{8.6}$$

and

$$t = \frac{b - \beta}{s_b} \tag{8.7}$$

where s_a and s_b are the sample standard deviations of a and b, respectively, have t-distributions both with $n - 2$ degrees of freedom. The sample variances s_a^2 and s_b^2 are given by

$$s_a^2 = s_{Y \cdot x}^2 \left(\frac{1}{n} + \frac{\bar{X}^2}{(n - 1)s_X^2} \right) \tag{8.8}$$

and

$$s_b^2 = s_{Y \cdot x}^2 / (n - 1)s_X^2 \tag{8.9}$$

where $s_{Y \cdot x}^2$ is given by (8.4) and s_X^2 is the sample variance of X.

A simple but common problem in the regression analysis is that of determining whether the slope β of a true regression line is zero. In other words, we want to know whether Y is linearly related to X. The hypothesis can be formulated as

$$H_0: \beta = 0.$$

In order to test the hypothesis H_0, substituting (8.9) into (8.7), we obtain

$$t = \frac{b}{s_b} = \frac{\sqrt{n - 1}\, b s_X}{s_{Y \cdot x}} \tag{8.10}$$

which has a t-distribution with $n - 2$ degrees of freedom.

At the α level of significance, a two-sided test rejects the H_0 if $t < -t_{\alpha/2}$ or $t > t_{\alpha/2}$ where $t_{\alpha/2}$ is the $100\alpha/2\%$ point of a t-distribution with $n - 2$ degrees of freedom. If the alternative is $H_1: \beta > 0$, we reject H_0 if $t > t_{\alpha}$, and if it is $H_1: \beta < 0$, we reject H_0 when $t < -t_{\alpha}$.

It is easy to see that from (8.6) and (8.7) the $100(1 - \alpha)\%$ confidence intervals for α and β are obtained as follows:

$$\alpha: \quad a \pm t_{\alpha/2}s_a$$
$$\beta: \quad b \pm t_{\alpha/2}s_b$$

where s_a and s_b are calculated from (8.8) and (8.9).

The use of the methods described in this section is illustrated next using an example.

Example 8.2 Table 8.1 presents the age (in years) and the crown length (in mm) of the maxillary deciduous central right incisor, one of the upper cutting teeth, of 15 *Macaca mulatta* monkeys. The data was furnished by Dr. Mildred Trotter, Washington University Department of Anatomy. It would be interesting to see if the age (X) of monkeys can be used to estimate the crown length (Y) based on a linear regression.

TABLE 5.1 Crown length of central incisor and age of 15 monkeys

Age	Crown length	Age	Crown length	Age	Crown length
0.06	5.05	0.36	5.25	1.77	3.35
0.08	4.95	0.56	4.40	1.94	3.40
0.08	5.10	0.83	5.15	2.14	3.05
0.18	4.85	1.02	4.30	2.44	2.65
0.28	4.40	1.23	4.15	2.44	3.00

The least square regression line of crown length on age is calculated using (8.2) and (8.3):

$$\hat{Y} = 5.162 - 0.934X.$$

We also calculated $s_Y = 0.892$ and $s_X = 0.901$ from which the standard deviation of Y given X is computed to be $s_{Y \cdot X} = 0.307$ applying (8.5). Now assuming a normal distribution for the Y for a given value of X, we would like to determine whether the true slope β can be considered nonzero on the basis of the data. In other words, we wish to test the hypothesis $H_0: \beta = 0$. From (8.9), we calculate

$$s_b = \frac{(0.307)}{\sqrt{14(0.901)}} = 0.091.$$

Using (8.10), we have

$$t = \frac{-0.934}{0.091} = -10.26.$$

With 13 degrees of freedom, the result is highly significant ($P < 0.0005$). We can conclude that the slope is significantly different from zero and that the age is a useful variable in estimating the crown length of monkeys. It is most interesting to note that the slope is negative, which implies that as monkeys get older, the crown length of teeth get shorter which is obvious on inspecting the wear on individual teeth. The average tooth wear is about 0.934 mm in length per year.

The 95% confidence interval for the slope β is obtained as

$$-0.934 \pm (2.16)(0.091),$$

or -1.131 to -0.737.

Some readers may think that it is more useful to find the regression equation of age on the crown length, which can then be used to estimate the age of monkeys by measuring the crown length. The regression line must be recalculated from (8.2) and (8.3) where X now denotes the crown length and Y the age. Readers may wish to verify that the new regression equation is

$$\hat{Y} = 5.028 - 0.952X$$

which can be used to estimate the age (Y) of monkeys based on the crown length (X). ▲

It is worthwhile at this point to make several remarks on the limitations and the interpretation of regression analysis. The following remarks apply to all analyses dealing with linear regression in general.

(a) A regression line as a prediction tool is, in general, applicable only within the range covered by the values of the predictor variable unless the straight line relation is known to extend beyond the range. In the first regression line of Example 8.2, the Y intercept is calculated to be $a = 5.16$, which means that the estimated mean crown length is 5.16 mm when the age is zero. The result is quite anomalous because teeth of newborn monkeys do not have measurable crown lengths. Measurements were made only of teeth which had completely emerged; therefore, the linear regression equation cannot be used for the growth phase. Often we may be dealing with a "small" linear segment of the regression model in which the overall relationship may actually be nonlinear.

(b) Given that the error terms ϵ_i's associated with the variable Y_i have the equal variance and are independent, deviation from normality in the distribution of ϵ usually has no serious consequence in the analysis of Sec. 8.3 particularly if n is large. This comment applies also to the methods to be described in the next section.

(c) A poor fit reflected by large $s_{Y \cdot X}$ or nonrejection of the hypothesis $H_0: \beta = 0$ may occur if a linear model is fitted to a nonlinear relationship. In this case, the deviation of Y_i's from the regression line reflects not only the true error term but also the deviation of a true curvilinear model from an assumed linear model. Conversely, however, the significance of the regression coefficient β does not necessarily mean that the line is the best fit. In this regard, a scatter diagram is useful in assessing whether the line is an appropriate model before any computations are made. The nonlinear regression model is discussed in Sec. 8.8.

(d) The last point to be made is that, as illustrated in Example 8.2, the equation $\hat{Y} = a + bX$ cannot be used directly in the form $\hat{X} = (Y - a)/b$ to obtain a regression line of X on Y. It must be recalculated using (8.2) and (8.3) after interchanging X and Y.

8.4 Inferences about Predicted Value

As it has already been stated, the least square regression line can be used to predict the mean value of Y corresponding to a given X, say X_0, without making any assumptions as to the distribution of Y. The predicted mean is

$$\hat{\mu}_{Y \cdot X_0} = a + bX_0,$$

where X_0 is not necessarily one of the X value used in computation of a and b.

In order to obtain a confidence interval for the mean value $\mu_{Y \cdot X_0}$, we need to assume that the distribution of ϵ is normal with a constant variance for all Y, as was the case in Sec. 8.3. The $100(1 - \alpha)\%$ confidence interval is given by

$$a + bX_0 \pm t_{\alpha/2}s_{Y \cdot X}\sqrt{\frac{1}{n} + \frac{(X_0 - \bar{X})^2}{(n - 1)s_X^2}} \tag{8.11}$$

where $t_{\alpha/2}$ denotes the upper $100\alpha/2\%$ point of the t-distribution with $n - 1$ degrees of freedom.

It is often the case in applications, however, that a confidence statement about the mean value is unimportant, whereas interest centers on a new single individual Y for given X_0. As has been noted already, the predicted $\hat{Y}_0$ for given X_0 is the same as above, namely, $\hat{Y}_0 = a + bX_0$. The confidence interval for an individual Y is better known as a *prediction interval*. The $100(1 - \alpha)\%$ prediction interval for an individual Y corresponding to X_0 is

$$a + bX_0 \pm t_{\alpha/2}s_{Y \cdot X}\sqrt{1 + \frac{1}{n} + \frac{(X_0 - \bar{X})^2}{(n - 1)s_X^2}}. \tag{8.12}$$

The only difference between (8.11) and (8.12) is the additional term "1" within the square root sign in (8.12). In order to avoid any misunderstanding, the interval (8.12) represents the limits between which we are $100(1 - \alpha)\%$ confident that a new single Y corresponding to the specified X_0 will lie, whereas the interval (8.11) represents the limits between which we are $100(1 - \alpha)\%$ confident that the true mean value of Y corresponding to X_0 will lie. Frequently, the prediction interval is incorrectly used as the limits which contain $100(1 - \alpha)\%$ of the population.

Example 8.3 Consider that we wish to estimate the systolic blood pressure of a five-year-old child. Based on the data for Example 8.1, the best linear estimate of the systolic pressure of the child is obtained by substituting 5 for X in the regression equation calculated in Example 8.1:

$$\hat{Y}_0 = 96.80 + (1.11)(5) = 102.4.$$

Now suppose we are interested in finding the 95% confidence interval for the mean systolic pressure of the population of five-year-old children. Using (8.11), we have

$$102.4 \pm (2.306)(20.99)\sqrt{0.1 + 0.011},$$

or 86.2 to 118.5. The corresponding confidence interval or prediction interval for the blood pressure of a randomly chosen five-year-old child is

$$102.4 \pm (2.306)(20.99)\sqrt{1 + 0.1 + 0.011}, \text{ or } 51.3 \text{ to } 153.4. \quad \blacktriangle$$

8.5 Correlation Coefficient

In the regression analysis, it is assumed that X is fixed while Y is random. Now suppose that both X and Y are continuous random variables (to be precise, they have a bivariate distribution) and we wish to measure the degree of relationship between X and Y. One such measure is a *simple linear correlation coefficient*, or *correlation coefficient*, to be short, and its population value is denoted by ρ.

Given a sample of n pairs (X_1, Y_1), (X_2, Y_2), $\ldots$, (X_n, Y_n), the sample correlation coefficient denoted by r which estimates ρ is computed as

$$r = \frac{\sum (X_i - \bar{X})(Y_i - \bar{Y})}{\sqrt{\sum (X_i - \bar{X})^2 \sum (Y_i - \bar{Y})^2}},$$

or

$$r = \frac{\sum X_i Y_i - n\bar{X}\bar{Y}}{(n-1)s_X s_Y}. \tag{8.13}$$

If the regression coefficient b (in $\hat{Y} = a + bX$) is already known, the coefficient r can be calculated by the formula:

$$r = b\frac{s_X}{s_Y}. \tag{8.14}$$

The correlation coefficient r is also known as *Pearson's correlation coefficient* after Pearson who defined it, or as the *product moment correlation coefficient*.

Equation (8.14) shows that the regression and correlation coefficients are related, and have the same sign. In practical applications, the choice between regression and correlation analyses depends on the purpose of the analysis.

The scatter diagrams associated with various values of the sample correlation coefficient are shown in Fig. 8.4. The larger the absolute value $|r|$, the greater the strength of linear association between X and Y. With experience one may achieve an intuitive feeling for the closeness of the relationship associated with a value of r.

It can be shown that the coefficient r is independent of the scale of measurement used for X and Y. In addition, the value of r satisfies the inequalities

$$-1 \leq r \leq 1,$$

and it will be plus 1 or minus 1 if, and only if, all points of the scatter diagram lie on a straight line with nonzero slope.

As with any other sample value such as the sample mean or the sample standard deviation, the reliability of r depends on the sample size. However,

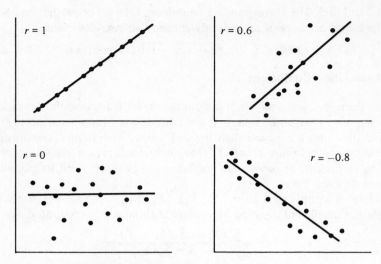

Fig. 8.4   Scatter diagrams and their values of r

even if the sample sizes are the same, values of r equal to 0.4 and 0.8, for example, only suggest that there are two positive correlations, one somewhat stronger than the other, and it is wrong to state that $r = 0.8$ is a relationship twice as good as that indicated by $r = 0.4$. In this regard, an objective numerical interpretation of the correlation coefficient would be useful. A more interpretable measure than r itself is given by:

Coefficient of determination $= 100r^2$.

Expressing (8.5) in terms of b^2 and substituting the result into (8.14), letting $n - 1 \approx n - 2$, we get

$$r^2 \approx \frac{s_Y^2 - s_{Y \cdot X}^2}{s_Y^2}. \tag{8.15}$$

Recall from Sec. 8.3 that $s_{Y \cdot X}^2$ estimates the variance of Y about the regression line: it is the variance of the error term ϵ and represents the variation in Y that is not accounted for by the linear relation of Y with X. Equation (8.15) may be written as

$$r^2 \approx \frac{\text{total variance of } Y - \text{unexplainable variance of } Y}{\text{total variance of } Y}. \tag{8.16}$$

Therefore, it is seen that the coefficient of determination may be interpreted as the percent of the total sample variation of Y that is accounted for by its linear association with X. For example, a correlation of 0.8 or -0.8 means that 64 % of the sample variation of the variable Y is accounted for by linear association with the variable X.

As has been stated, r is used to estimate the population correlation coefficient ρ. Depending on the value of ρ, the following statement is made:

(a) $\rho > 0$: X and Y are positively correlated;
(b) $\rho = 0$: X and Y are not correlated;
(c) $\rho < 0$: X and Y are negatively correlated.

It should be emphasized again that the correlation coefficient provides a measure of linear relationship only. Thus, X and Y can be functionally related, but ρ can be small or zero. A simple example is when Y and X are perfectly associated by the relation, $Y = \sin X$ where $0 \leq X \leq 2\pi$.

8.6 Inference about Correlation Coefficient

A simple but most common problem in correlation analysis is testing the hypothesis of a zero correlation coefficient formulated by

$$H_0 : \rho = 0.$$

Let r denote the sample correlation coefficient based on a sample size of n. To test the hypothesis of no correlation, we use the variable

$$t = r\sqrt{\frac{n-2}{1-r^2}} \tag{8.17}$$

which has a t-distribution with $n - 2$ degrees of freedom provided that either X or Y or both variables are normally distributed.

At the α level of significance, against the alternative $H_1 : \rho \neq 0$, a test is made by rejecting H_0 when either $t < -t_{\alpha/2}$ or $t > t_{\alpha/2}$, where $t_{\alpha/2}$ is the $100\alpha/2\%$ point of a t-distribution with $n - 2$ degrees of freedom. We can state that the correlation between X and Y is significantly different from zero when the H_0 is rejected against the H_1. Against the $H_1 : \rho > 0$, we reject H_0 if $t > t_{\alpha}$, and against $H_1 : \rho < 0$, we reject H_0 if $t < -t_{\alpha}$. Incidentally, (8.10) and (8.17) are algebraically equivalent. Thus, the two hypotheses $H_0 : \beta = 0$ and $H_0 : \rho = 0$ are equivalent, and there is no need for performing both tests. A convenient way of testing the hypothesis $H_0 : \rho = 0$ is by comparing r with critical values tabulated in Table A9.

When the joint distribution of X and Y is a bivariate normal, a very useful transformation of the sample correlation coefficient r is given by

$$Z(r) = \frac{1}{2} \log_e \frac{1+r}{1-r}. \tag{8.18}$$

If the corresponding population correlation is ρ, then $Z(r)$ has an approximately normal distribution with the mean and variance given by

$$Z(\rho) = \frac{1}{2} \log_e \frac{1+\rho}{1-\rho}$$

and

$$\text{Var}\,[Z(r)] = \frac{1}{n - 3},$$

respectively. This approximation is fairly good as long as $n \geq 10$.

Values of the transformed correlation coefficient, $Z(r)$, are given in Appendix Table A14.

We shall now describe the three common applications of the above transformation.

(a) Suppose we wish to test the hypothesis $\rho = \rho_0$ where ρ_0 is a value between -1 to 1. [Note that (8.17) is restricted for testing $H_0: \rho = 0$.] We compute r and $Z(r)$ using (8.18). Let

$$Z(\rho_0) = \frac{1}{2} \log_e \frac{1 + \rho_0}{1 - \rho_0}.$$

The test criterion is then given by the variable

$$Z = \frac{Z(r) - Z(\rho_0)}{1/\sqrt{n - 3}}$$

which has approximately the standard normal distribution if $\rho = \rho_0$. For example, against a two-sided alternative at the 5% significance level, we reject $H_0: \rho = \rho_0$ if $Z > 1.96$ or $Z < -1.96$.

(b) Suppose that a sample of n_1 from one population and an independent sample of n_2 from a second population are obtained. How do we test the hypothesis that the two correlation coefficients are equal, $\rho_1 = \rho_2$? Let r_1 and r_2 be the two sample correlation coefficients. We calculate $Z(r_1)$ and $Z(r_2)$, and as a test statistic, we use

$$Z = \frac{Z(r_1) - Z(r_2)}{\sqrt{1/(n_1 - 3) + 1/(n_2 - 3)}}$$

which has approximately the standard normal distribution. Hence, the hypothesis of a common correlation coefficient can be tested by referring to Appendix Table A1. It should be remembered that this test is valid only when r_1 and r_2 are calculated from two independent samples.

As a simple illustration, let $r_1 = 0.5$, $r_2 = 0.8$, $n_1 = 20$, and $n_2 = 40$. We calculate $Z = -1.88$ which gives the two-sided P value of 0.1. The evidence does not suggest the rejection of a common correlation coefficient of the two populations.

(c) Finally, consider the problem of calculating the confidence interval for ρ, based on a sample of n observations. Since $Z(r)$ is approximately normally distributed with mean $Z(\rho)$ and variance $1/(n - 3)$, we can write

$$P(Z(r) - z_{\alpha/2}/\sqrt{n - 3} < Z(\rho) < Z(r) + z_{\alpha/2}/\sqrt{n - 3}) = 1 - \alpha$$

where $z_{\alpha/2}$ represents the upper $100\alpha/2\%$ of the standard normal distribution. This equation has to be manipulated to obtain the bounds for ρ. Although we will not go through the mathematics, the approximate $100(1 - \alpha)\%$ confidence interval for ρ can be obtained as

$$\tanh[Z(r) \pm z_{\alpha/2}/\sqrt{n - 3}]$$

where $\tanh(x)$ denotes the hyperbolic tangent of x and can be calculated as $\tanh(x) = (e^x - e^{-x})/(e^x + e^{-x})$. An easier method of obtaining the confidence interval is to simply read the interval from Appendix Fig. A10 or A11. An interpolation may be used in practice.

Example 8.4 (a) Using the data of Example 8.1, test whether the systolic blood pressure is correlated with the diastolic blood pressure. (b) For the same data, test the hypothesis that the true correlation coefficient is equal to 0.8. (c) Determine the 95 % confidence interval for the population correlation coefficient ρ.

First, using (8.13), the sample correlation coefficient is calculated to be $r = 0.763$. The hypothesis for the first question is that the correlation coefficient ρ is zero. To test this hypothesis, we apply (8.17), which yields

$$t = 0.763\sqrt{\frac{10 - 2}{1 - 0.763^2}} = 3.34.$$

From Table A2 it will be found that the P value corresponding to $t = 3.34$ with 8 degrees of freedom is less than 0.05. We, therefore, reject the hypothesis and state that the correlation between the systolic and diastolic blood pressure is significant at $P < 0.05$. Alternatively, the test can be made by using Table A9. From Table A9, the critical value at $P = 0.05$ when $n = 10$ is found to be 0.632. Since $0.763 > 0.632$, the hypothesis of zero correlation is rejected at $P < 0.05$.

To test the hypothesis $\rho = 0.8$, we have to use the method based on the transformation given by (8.18). From Table A14 we get $Z(0.763) = 1.00$ using an interpolation and $Z(0.8) = 1.10$. The test value is calculated as

$$Z = \frac{1.00 - 1.10}{1/\sqrt{7}} = -0.265.$$

Clearly, the hypothesis $\rho = 0.8$ cannot be rejected.

Finally, the 95 % confidence interval for ρ is obtained from Fig. A10. For the above data, the 95 % confidence interval is 0.22 to 0.93. ▲

It is worthwhile at this point to remark on the reliability and limitations of the correlation coefficient. (a) When the sample size is small, the reliability of r as an estimate of ρ is low due to the large sample variance of r. The point is illustrated in part by the wide confidence interval of Example 8.4. (b) As has been already stated, the correlation coefficient is only a measure of a linear relationship between two variables. However, a nonzero correlation coefficient does not necessarily imply that the relation is strictly linear. (c)

The relationship given by the correlation coefficient is mathematical and may be devoid of any cause or effect implications.

Quite often a high correlation exists because two variables are both related to other factors. The following two humorous stories illustrate this point.

As the first example, consider the correlation coefficient between teachers' salaries and the price of liquor based on data between 1961 to 1977 in the United States. The correlation coefficient was found to be very high. During the period there was a steady rise in wages to compensate for inflation and a continuous rise in liquor prices at about the same pace. Clearly, it is absurd to implicate any cause or effect relation between the two variables. The second example concerns a high positive correlation over a period of years in the late 1800's between the number of storks sighted and the number of births in a European city. The fact is that many of the homes in that area at that time were built with thatched roofs which are ideal for nesting storks. As the city grew, more families built homes, and there were more nesting places. And, of course, more families produced more babies. Thus, the high correlation merely reflected the common effect of the upward trend of the two variables in time.

The moral of these stories is that "correlation does not mean causation" and the correct interpretation of a correlation coefficient requires familiarity with the field of application in addition to statistics. Sometimes we ask, "Are the two variables related in some way?" and "If they are related, is there any reason they should be?" The first question may be answered using statistics, but the second question cannot.

8.7 Analyses Involving More Than One X

In this section we shall be concerned with some simple statistical procedures useful for analyzing data that consist of more than two measurements on each member of individuals or objects.

8.7.1 Correlation Matrix

Suppose that k, $k > 1$, measurements were made on each of n individuals in a sample. One convenient way of simultaneously presenting the linear interrelations between the k variables is to calculate the *sample correlation matrix* denoted by **R**. It is a $k \times k$ square matrix with the element r_{ij} which is the sample correlation coefficient between the ith and jth variables calculated by (8.13):

$$\mathbf{R} = \begin{bmatrix} 1 & r_{12} & r_{13} & \cdot & \cdot & r_{1k} \\ r_{21} & 1 & r_{22} & \cdot & \cdot & r_{2k} \\ \cdot & \cdot & \cdot & & & \\ r_{k1} & r_{k2} & r_{k3} & \cdot & \cdot & 1 \end{bmatrix}$$

The matrix **R** is symmetric since $r_{ij} = r_{ji}$ for all i and j, and its diagonal elements are all unity because of unit correlation coefficient of each variable with itself. Usually, the correlation matrix is calculated using a computer, and many programs are avilable for the purpose. Not only does the correlation matrix enable us to examine the correlation between many pairs of variables simultaneously, but it also provides a basis for further analyses which will not be dealt with in this book because of their complexity.

8.7.2 Partial Correlation Coefficient

Consider any three selected variables X_i, X_j, and X_h among $k, k \geq 3$ variables and let r_{ij}, r_{ih}, and r_{jh} be the correlation coefficients between X_i and X_j, between X_i and X_h, and between X_j and X_h calculated from the same sample. It is quite conceivable that some of the apparent correlation between X_i and X_j may be due to the association of each with X_h. The *partial correlation coefficient* between X_i and X_j holding X_h fixed, denoted by $r_{ij \cdot h}$, is the correlation between X_i and X_j after eliminating the common influence of X_h. To be a little more precise, it is the correlation between X_i and X_j adjusted for the linear regression of each on X_h.

Using the elements of the correlation matrix, the sample partial correlation coefficient $r_{ij \cdot h}$ can be computed as

$$r_{ij \cdot h} = \frac{r_{ij} - r_{ih} r_{jh}}{\sqrt{(1 - r_{ih}^2)(1 - r_{jh}^2)}}, \tag{8.19}$$

which estimates the analogously defined population partial correlation coefficient, $\rho_{ij \cdot h}$. Obviously, we can calculate $r_{ih \cdot j}$ or $r_{jh \cdot i}$ in an analogous way.

If either X_i or X_j or both variables are normally distributed it is possible to test that the true partial correlation coefficient is zero. The exact test criterion for $H_0: \rho_{ij \cdot h} = 0$ is given by a form similar to (8.17), namely,

$$t = r_{ij \cdot h} \sqrt{\frac{n - 3}{1 - r_{ij \cdot h}^2}} \tag{8.20}$$

which has a t-distribution with $n - 3$ degrees of freedom. The test is performed in the same way as for testing the hypothesis $H_0: \rho = 0$, the only difference being the degrees of freedom which is $n - 3$ instead of $n - 2$. The test can be facilitated by the critical values shown in Table A9 assuming the sample size is one less. That is, for example, if the sample size is 20, then we refer to Table A9 with $n = 19$.

The partial correlation coefficient can be extended to four or more variables in the same general way, although we shall not do that here. See Ref. 9.

Example 8.5 A correlation matrix of systolic blood pressure (X_1), diastolic pressure (X_2), height (X_3), and weight (X_4) is calculated from data based on a sample of 200 high school girls in the St. Louis area. Only the upper diagonal part of the correlation matrix is given below.

$$
\begin{array}{c}
\quad\quad X_1 \quad X_2 \quad X_3 \quad X_4 \\
\begin{array}{c} X_1 \\ X_2 \\ X_3 \\ X_4 \end{array}
\left[
\begin{array}{cccc}
1 & .427 & .146 & .375 \\
  & 1 & .091 & .171 \\
  &   & 1 & .416 \\
  &   &   & 1
\end{array}
\right]
\end{array}
$$

Consider the correlation coefficient of X_1 with X_3 and with X_4. From Table A9, r_{13} is significant at $P < 0.05$ and r_{14} is at $P < 0.001$. It is interesting to examine the partial correlation between X_1 and X_3 holding X_4 fixed since the significant correlation between the blood pressure and height might have been caused by the common influence of the weight. Using (8.19), remembering that $i = 1, j = 3$, and $h = 4$ in the present problem, we calculate

$$
r_{13 \cdot 4} = \frac{0.146 - (0.375)(0.416)}{\sqrt{(1 - 0.375^2)(1 - 0.416^2)}} = -0.012.
$$

Thus, the partial correlation between systolic blood pressure and height is rather small if weight is held fixed. From (8.20), we have

$$
t = -0.012 \sqrt{\frac{197}{1 - (-0.012)^2}} = -0.168,
$$

which is clearly not significant at 197 degrees of freedom. On the other hand, the partial correlation between blood pressure and weight with the effect of height held constant is calculated to be 0.349, and it can be shown that the correlation is significant at $P < 0.001$. Apparently, weight and blood pressure have a positive relation even in the population of young girls.

The ordinary and partial correlation coefficients in this example can be conveniently depicted in a diagram (Fig. 8.5 and Fig. 8.6), where a "two-way arrow" sign indicates the correlation.

Figure 8.5 illustrates the relationship indicated by the ordinary correlation coefficient between the three variables, and it suggests the relationship between blood pressure and height. The partial correlation analysis, however,

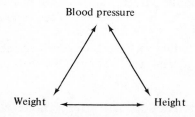

Fig. 8.5   Relationship indicated by ordinary correlation coefficients

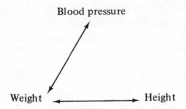

Fig. 8.6 Relationship indicated by partial correlation coefficients

shows that the latter correlation is the result of the weight which is related to both blood pressure and height. In other words, height is correlated with blood pressure through weight, and if the effect of the dependence on weight is removed, blood pressure no longer varies with height. The point is depicted in Fig. 8.6. ▲

Partial correlation can be a very useful tool in determining whether a relationship is spurious. A *spurious correlation* is defined as a relationship between two variables in which the correlation is solely the result of both variables varying along with some other primary variable. In spite of the usefulness of the partial correlation, like a simple correlation coefficient, the partial correlation coefficient is a purely mathematical index and is devoid of cause or effect implications. Also, as in the case of a simple correlation coefficient, the partial correlation reflects only the linear relation between two variables adjusted for the linear relation of each on the third variable.

8.7.3 Multiple Regression

We have considered in Sec. 8.2, 8.3, and 8.4, in detail, the simple linear regression model involving one fixed variable X, namely,

$$Y = \alpha + \beta X + \epsilon.$$

In many instances, it is not reasonable to assume that the observed values of Y are influenced by only one independent variable X. In many practical situations, therefore, more complex models are needed where information on more than one predictor variable is required to obtain a better relation and prediction of a particular dependent variable Y. For example, we might wish to estimate the yield of wheat, Y, for different amounts of phosphate, X_1, nitrogen, X_2, and potash, X_3, applied to the soil. The economist may wish to determine the inflation rate, Y, using money supply, X_1, capacity utilization rate, X_2, and excess of labor cost, X_3. In either case, we may wish to consider the regression model of the form:

$$Y = \alpha + \beta_1 X_1 + \beta_2 X_2 + \beta_3 X_3 + \epsilon$$

where Y, X_1, X_2, and X_3 are the variables defined as above depending on the problem and α, β_1, β_2, and β_3 are the coefficients to be determined.

More generally, suppose there are k fixed variables $X_1, X_2, \ldots, X_k$, and that the dependent variable Y is linearly related to the X's. As a natural extension to the simple linear regression model, we have a *multiple linear regression model:*

$$Y = \alpha + \beta_1 X_1 + \beta_2 X_2 + \cdots + \beta_k X_k + \epsilon$$

where ϵ represents the residual or error of Y not accounted for by the regression line. We postulate that the ϵ has an independent distribution with zero mean and homogeneous variance for each observation on Y. The primary purpose of the multiple regression is to determine a model of prediction or estimation for a particular variable, Y. In addition, the multiple regression is a useful tool in studying the relation between one dependent random variable and several other variables. An important point is that independent variables may be jointly explicatory of Y even if individually they may not be. The following artificial example illustrates such a situation. Although Y cannot be explained simply by either X_1 or X_2 alone, Y is precisely explainable by X_1 and X_2 if used jointly since $Y = -1000 + X_1 + X_2$.

Y	X_1	X_2
0	2	998
1	998	3
2	1	1001
3	1001	2
4	−1	1005
etc.		

Suppose we have n sets of observations. Then, for the purpose of analysis, the data can be arranged in the following form which may be called the *data matrix:*

$$\begin{bmatrix} Y_1 & X_{11} & X_{21} & \cdots & X_{k1} \\ Y_2 & X_{12} & X_{22} & \cdots & X_{k2} \\ \cdot & \cdot & \cdot & \cdots & \cdot \\ \cdot & \cdot & \cdot & \cdots & \cdot \\ \cdot & \cdot & \cdot & \cdots & \cdot \\ Y_n & X_{1n} & X_{2n} & \cdots & X_{kn} \end{bmatrix}$$

where X_{ij} represents the jth observation for the ith variable. Because of the same reason that we require at least three observations in the simple linear regression, we require that in multiple regression analysis

$$n > k + 1.$$

For estimating the coefficients α, $\beta_1, \beta_2, \ldots, \beta_3$, the Gauss-Markov theorem applies. That is, the least squares method provides the best unbiased estimators of the coefficients. The computation involved, however, is quite laborious, and nowadays it is almost always performed by a computer. Many computer programs are available for the multiple regression analysis (see Chap. 10).

When there are only two variables, Y and X, the simple correlation coefficient is used as a measure of the linear relation between them. How do we express the relationship between Y and X's? Analogous to the simple correlation is the *multiple correlation coefficient*. It is the linear correlation coefficient between Y and $\beta_1 X_1 + \beta_2 X_2 + \cdots + \beta_k X_k$. It may be interpreted as a correlation between Y and all other X's when X's are combined in the best possible linear manner to express Y. The multiple regression computer programs as a rule give the multiple correlation coefficient as a part of the output. The range of the multiple correlation coefficient is 0 to 1. Thus, the only statement we can make would be either Y is correlated to the X's or not correlated, but not positively or negatively. Let R denote the population multiple correlation coefficient and $\hat{R}$ be the corresponding sample coefficient. For testing the hypothesis $H_0 : R = 0$ when Y is assumed to be normal, we use the rejection region

$$F = \frac{(n - k - 1)\hat{R}^2}{k(1 - \hat{R}^2)} > F_\alpha \qquad (8.21)$$

where F_α denotes the upper $100\alpha\%$ point of the F-distribution with $(k, n - k - 1)$ degrees of freedom. This test is equivalent for testing the significance of a linear regression of Y on X's. In other words, if $H_0 : R = 0$ is rejected, we decide that $R > 0$ or, equivalently, the regression of Y on X's is significant. Note saying that a regression is significant implies that at least one $\beta_i \neq 0$.

As in the case of the simple regression, R^2 is called the *coefficient of determination*, and it represents the proportion of the total variance in Y explainable by its linear relation with $X_1, X_2, \ldots, X_k$. This interpretation of R^2 is reasonable since $\hat{R}^2$ can be expressed as

$$\hat{R}^2 = \frac{\sum (\hat{Y}_i - \bar{Y})^2}{\sum (Y_i - \bar{Y})^2}.$$

It is the ratio of the sum of squares due to regression to the total sum of squares. The larger it is, the better the fitted equation explains the variation in the data. The partition of the sum of squares is dealt with in detail in Sec. 9.7. Examples of the multiple regression are given in Sec. 9.7 and 10.4.

*8.8 Nonlinear Regression Model

So far we have only considered linear relationships in regression models. However, the regression model can be of a more complex type involving powers, roots, logs, or any other function of X's. For example, the relationship may be of the form $Y = \alpha + \beta_1 \sin(t) + \beta_2 \cos(t) + \epsilon$, where t represents the independent variable. Note that the relationship can be expressed as a standard multiple linear regression model by letting $X_1 = \sin(t)$ and

$X_2 = \cos(t)$. Thus, multiple linear regression techniques can be applied to analyze the problem. As the second example, consider the model

$$Y = \alpha X_1^{\beta_1} X_2^{\beta_2} \epsilon$$

where α, β_1, and β_3 are unknown parameters with ϵ representing the random error. Taking logarithms of both sides of the equation converts the model into a linear form

$$\log Y = \log \alpha + \beta_1 \log X_1 + \beta_2 \log X_2 + \log \epsilon.$$

This model too can be handled as the standard multiple linear regression model. The nonlinear regression model is said to be *intrinsically linear* if suitable transformation changes it into the standard multiple linear regression form. Many computer programs for multiple regression analysis have optional capability of handling the intrinsically linear model.

Another well-known model in regression analysis, which can be handled by a standard linear regression computer program, is the *polynomial regression*, which is of the form

$$Y = \alpha + B_1 X + B_2 X^2 + \cdots + \beta_p X^p + \epsilon$$

where p is some positive integer and α, $\beta_1, \ldots, \beta_p$ are the coefficients to be estimated.

Finally, there are many intrinsically nonlinear models which cannot be expressed in linear form. An example of such models is *asymptotic regression* of the following form although there are several different versions:

$$Y = \alpha + \beta \gamma^x + \epsilon$$

where α, β, and γ are unknown coefficients to be estimated. Note that the model cannot be treated as a multiple linear regression model. Several computer programs are available for fitting the data to the type of nonlinear model. See Ref. 1 and Ref. 9 for more detailed discussion of the nonlinear regression.

8.9 Contingency Coefficient

The correlation coefficient discussed in Sec. 8.5 serves as a measure of linear association between two continuous variables. Now consider the problem of indexing the strength of statistical association between qualitative or categorical attributes. Although many coefficients of association for such purposes have been proposed, no one coefficient exists to everyone's real satisfaction. First, the value of χ^2 computed from contingency tables is not a suitable measure of association, since its upper limit is infinite as the sample size increases. We shall give one simple coefficient known as *contingency coefficient*.

To compute the contingency coefficient between two sets of categories, say $A_1, A_2, \ldots, A_r$ and $B_1, B_2, \ldots, B_c$, we arrange the data in the form of the contingency table like Table 6.5. Let χ^2 be the value of the chi-square computed. Then the contingency coefficient is defined by

$$C = \sqrt{\frac{\chi^2}{n + \chi^2}}$$

where n denotes the total sample size.

In general, it would be desirable for correlation coefficients to have at least the following characteristics: (a) the coefficient should be zero when there is a complete absence of any association, and (b) the coefficient should equal unity when the variables are completely dependent. The contingency coefficient, C, has the first but not the second property. The upper limit for C depends on the size of the contingency table. The attainable upper bound for C turns out to be

$$\sqrt{\frac{\min(r - 1, c - 1)}{1 + \min(r - 1, c - 1)}}.$$

Thus, for example, the upper limit for C in the case of a three-by-three contingency table is $\sqrt{2/3} = 0.816$. The fact that the upper limit of C depends on the sizes of the contingency table is the second limitation of C. Two contingency coefficients are not comparable unless they are obtained from contingency tables of the same size.

The third limitation may be that C, unlike the correlation coefficient, is nonnegative even if attributes are negatively associated.

The test of hypothesis of association between qualitative or categorical attributes can be, of course, carried out using the chi-square test described in Chap. 6.

Example 8.6 Consider the data presented in Table 6.3 for which the value of χ^2 was calculated to be 13.3 (see Example 6.5). The contingency coefficient for the data is

$$C = \sqrt{\frac{13.3}{50 + 13.3}} = 0.46.$$

The attainable upper bound of C for a two-by-two table is 0.707. ▲

8.10 Biserial Correlation

The problem of biserial correlation arises when we are concerned with the relation between a continuous variable X and a dichotomized variable Y which can take two values only. We can assume these values to be 1 and 0.

Let n_1 and n_2 be the number of 1's and 0's, respectively, in a sample of Y, and p be the proportion of 1's, that is, $p = n_1/(n_1 + n_2)$. We further represent by $\bar{X}_0$ and $\bar{X}_1$ the means of X in the groups for $Y = 0$ and $Y = 1$, respectively. The *sample biserial correlation coefficient* which can be used as a measure of relationship between X and Y is given by

$$r = \frac{n(\bar{X}_1 - \bar{X}_0)\sqrt{pq}}{(n-1)s_X} \tag{8.22}$$

where s_X is the sample standard deviation, computed using all X's and $q = 1 - p$.

It is worthwhile to make a few comments about this coefficient. (a) Actually, the regular formula for the correlation coefficient given by (8.13) reduces to (8.22) when Y is dichotomized. (b) The value of r is related to that of the t-statistic for comparing the means of two normal populations in the following way:

$$\frac{r^2}{1-r^2} = \frac{t^2}{n_1 + n_2 - 2}. \tag{8.23}$$

Suppose it is desired to test the hypothesis of no correlation between X and Y variables. Clearly, this hypothesis is related to the question of whether the means of two parts of X corresponding to $Y = 0$ and $Y = 1$ are equal or not. In fact, because of the one-to-one relation given by (8.23), the hypothesis can be tested by the t-test. If the two parts of X, corresponding to $Y = 0$ and $Y = 1$, have a common variance, the test of no correlation can be performed using the following t-statistic which has $n_1 + n_2 - 2$ degrees of freedom:

$$t = \frac{\bar{X}_1 - \bar{X}_0}{s_p\sqrt{\dfrac{1}{n_1} + \dfrac{1}{n_2}}} \tag{8.24}$$

where s_p is the pooled standard deviation. The application of the formula (8.24) was described in Sec. 7.5.

Example 8.7 The closing wholesale prices of a certain commodity were recorded over 30 different days along with the weather conditions for each day. Nine days were cloudy, and 21 days were classified as fair. The sample means and variances of the prices grouped for the weather conditions were:

21 fair days	9 cloudy days
$\bar{X} = \$41.52$	$\bar{X} = \$40.36$
$s^2 = 1.42$	$s^2 = 1.57$

Is the price correlated with the weather conditions? We have

$$s_p^2 = \frac{(21-1)1.42 + (9-1)1.57}{30 - 2} = 1.46,$$

and using (8.24),

$$t = \frac{41.52 - 40.36}{1.21\sqrt{1/21 + 1/9}} = 2.41.$$

With 28 degrees of freedom, the value is significant at $P < 0.05$, and thus we can state that the relation between the prices and the weather conditions were significant. ▲

8.11 Rank Correlation

Consider the following seven paired observations. The relationship between X and Y should be a perfect one since $Y = \log X$.

$$
\begin{array}{llllll}
X: & 0.1 & 0.5 & 0.7 & 1 & 2 & 10 \\
Y: & \log 0.1 & \log 0.5 & \log 0.7 & 0 & \log 2 & \log 10
\end{array}
$$

The correlation coefficient, however, is calculated to be $r = 0.845$ instead of one. The restriction of the correlation coefficient, r, given by (8.13) is that (a) it measures only a degree of a linear relation and (b) the inference on parameters requires that the two variables are normally distributed. It would be desirable to have a more general measure of correlation and to have a test that requires no assumption on distribution.

There are several statistics available that are based on ranks and called *rank correlation coefficients*. They are nonparametric analogues of the ordinary correlation coefficient. We shall consider one of these, Spearman's rank correlation coefficient, r_s. As its name implies, it is simply the ordinary correlation coefficient calculated for the ranks of the variables rather than for the numerical values. Let D_i denote the difference between the ranks of the ith pair, after X's and Y's are ranked separately among themselves. A computational form for r_s is given by

$$r_s = 1 - \frac{6 \sum D_i^2}{n(n^2 - 1)} \tag{8.25}$$

where n is the number of pairs. Tied values are assigned the average of the ranks that would have been given had the values been slightly different. The ranking of observations can be made from low to high or high to low as long as it is done the same way for each variable.

The rank correlation coefficient, like the ordinary correlation coefficient, may take on values from -1 to 1. A value of -1 or 1 means perfect monotone relation between variables. Thus, for example, if X and Y are related by $Y = X^2$ or $Y = \log X$ where $X > 0$, then $r_s = 1$, as one can verify numerically using the above seven pairs of observations in the case of $Y = \log X$.

The hypothesis of the independence of the two variables can be tested by an approximate method if the sample size is large. The test statistic has a form identical to that using r:

$$t = r_s \sqrt{\frac{n-2}{1 - r_s^2}}, \tag{8.26}$$

which has an approximate t-distribution with $n - 2$ degrees of freedom if n is at least greater than about 15.

In the case when n is not large, Table A12 of the appendix gives two-sided critical values of r_s at several significance levels. For a one-sided test of zero correlation against positive (or negative) correlation, the significance level should be halved.

Example 8.8 In a test of fine weight discrimination, two individuals A and B each ranked ten small objects in order of their judged weight. The results are shown in the following table, in the first three columns. Did the two individuals tend to agree?

The hypothesis is that the correlation is zero, in this case that the degree of agreement between the two individuals is not significant.

Object	A	B	D_i	D_i^2
1	5	6	1	1
2	6	8	2	4
3	8	5	3	9
4	2	3	1	1
5	9	7	2	4
6	7	9	2	4
7	3	1	2	4
8	10	10	0	0
9	1	4	3	9
10	4	2	2	4

Substituting $\sum D_i^2 = 40$ into (8.25), we calculate

$$r_s = 1 - \frac{6(40)}{10(100 - 1)} = 0.758.$$

At $\alpha = 0.05$ level of significance, we reject the hypothesis of zero correlation, and thus, the agreement between the two is determined to be significant. ▲

It is worth noting that the rank correlation coefficient, although it may be a more general measure than the ordinary correlation coefficient, is by no means a perfect coefficient of relationship. For example, consider five pairs of observations $(-2, 4)$, $(-1, 1)$, $(0, 0)$, $(1, 1)$, and $(2, 4)$. Clearly, X and Y are perfectly related since $Y = X^2$. However, it can be calculated that $r_s = 0.05$. This, of course, is due to the fact that the r_s is only a measure of any monotonic relationship between X and Y. It should also be emphasized that in case of the bivariate normal distribution the test given by Eq. (8.17) using r is more efficient than that by Eq. (8.26). For the normal distribution with population correlation ρ, it is known that the rank correlation coefficient r_s tends to be smaller than r and underestimates ρ.

8.12 Summary

The methods of regression and correlation analyses are described in this chapter. Some of the simple but important formulas frequently encountered in applications are summarized below.

Regression coefficient:

$$a = \bar{Y} - b\bar{X},$$

$$b = \frac{\sum X_i Y_i - n\bar{X}\bar{Y}}{(n-1)s_X^2} = r\frac{s_Y}{s_X}.$$

Standard error of regression line:

$$s_{Y \cdot X} = \sqrt{\frac{n-1}{n-2}(s_Y^2 - b^2 s_X^2)}.$$

Correlation coefficient:

$$r = \frac{\sum X_i Y_i - n\bar{X}\bar{Y}}{(n-1)s_X s_Y} = b\frac{s_X}{s_Y}.$$

Test for zero correlation or zero slope of regression line:

$$t = r\sqrt{\frac{n-2}{1-r^2}} = \frac{\sqrt{n-1}\,b s_X}{s_{Y \cdot X}} \qquad (\text{d. f.} = n-2).$$

PROBLEMS

1. An aptitude test score (X) on five randomly selected applicants and a measure of the productivity (Y) of each individual are examined by a management in a study of a reliability of the aptitude test.

X	Y	X	Y	X	Y
20	29	17	35	20	38
19	33	23	32		

(a) Find the equation of the line that can be used to predict productivity from the test score.
(b) Calculate the standard error of the line.
(c) Test the hypothesis $H_0: \beta = 0$ against $H_1: \beta > 0$.
(d) Is the aptitude test useful in predicting productivity?

2. The following is a sample of the weekly wages, X, of workers in a certain industry and the weekly expenditure, Y, for each individual for entertainment. (The figures are in dollars.)

X	Y	X	Y	X	Y
171	7.5	270	26.7	279	21.0
207	14.0	180	8.4	255	19.0
234	20.1	216	15.0	225	12.8

(a) Find the regression line of Y on X.
(b) Calculate the standard error of the line.
(c) Test the hypothesis $H_0: \beta = 0$ against $H_1: \beta \neq 0$.
(d) Are you able to state that the income and expenditure are significantly related?
(e) Determine the 90% confidence interval for α and that for β.
(f) What is the expected expenditure of a worker making $200 per week?
(g) Determine the 90% confidence interval for the expected expenditure of the worker in (f).

3. What would you guess the value of the correlation coefficient for the following pairs of variables to be?

(a) The amount of education and income.
(b) The weights of monozygotic twins.
(c) Statistics grade and weight.
(d) The prices of used cars and their ages.
(e) *IQ* level and brain size.

4. A correlation coefficient of 0.50 is found from a sample of 28 pairs.

(a) Can it be regarded as significantly different from zero on the basis of the 5% level of significance?
(b) Construct the 99% confidence interval for the population correlation coefficient.

5. A sample of size 52 is taken from a population with the true correlation coefficient of 0.6.

(a) Approximate the probability that the sample correlation coefficient is 0.4 or less.
(b) And 0.8 or above.

6. Based on 15 randomly selected American couples, the correlation coefficient between heights of husband and wife was 0.89. What conclusion could you draw from the number?

7. In a public opinion survey the following question was asked: do you

think massage parlors should be placed under legal ban? In one district
the results summarized according to age were as follows:

	For ban	Against
40 or under	17	62
Over 40	45	22

Does this provide good evidence of a correlation between age and opinion on this issue?

8. The following data show the mean water temperature in February
   and March for the last eight years versus the mean number of days
   after January 1 that salmon migrate downstream for the corresponding
   year.

Year	1	2	3	4	5	6	7	8
Temperature (°C)	3.3	2.5	5.6	3.8	3.4	5.7	3.9	2.6
Days to migration	117	110	95	112	112	95	117	117

Predict the mean migration data for a year in which the mean February
and March water temperature is 3.0°C. Also give the 95% confidence
interval for this prediction.

9. A study was performed to determine the relationship between the
   tensile strength (X) in 1000 psi and Brinell hardness (Y) based on 16
   copper specimens. The observations yielded the following results:

   $$\bar{X} = 37.4, \qquad \bar{Y} = 103.9, \qquad s_X = 1.77, \qquad s_Y = 3.06,$$
   $$\sum (X_i - \bar{X})Y_i = 28.32.$$

   (a) Assuming normality, test the hypothesis that the true correlation
       coefficient ρ between the tensile strength and hardness is zero.
   (b) Find the 95% confidence interval for ρ.

10. The following data show the weight (in lb) and glucose level (in
    mg/100 ml blood) of 15 healthy adult males. Test whether the two variables are correlated. It has been suggested that the weight and glucose
    level are positively correlated.

Weight	Glucose	Weight	Glucose	Weight	Glucose
165	108	140	107	170	86
173	84	210	120	160	103
205	106	180	100	167	104
130	78	185	107	168	98
181	101	160	103	141	101

11. Let r_{XY} be the sample correlation coefficient between X and Y and
    r_{ZW} the sample correlation coefficient between Z and W based on the

transformation:

$$Z_i = aX_i + b,$$
$$W_i = cY_i + d$$

where a, b, c, and d are any positive constants. Show that

$$r_{ZW} = r_{XY}.$$

12. In a sample of 220 high school boys, the correlation coefficient between systolic blood pressure and weight is found to be 0.51. The corresponding coefficient calculated from 200 high school girls was 0.38. (See Example 8.5.) Is the difference between the two correlation coefficients significant?

*13. (a) Show that the approximate two-sided test for the hypothesis H_0: $\rho = 0$ with a significance level α is given by rejecting the H_0 when

$$|\sqrt{n-3}\, Z(r)| > z_{\alpha/2}$$

where $Z(r)$ is the transformation defined in Eq. (8.18) and $z_{\alpha/2}$ denotes the upper $100\alpha/2\%$ point of the standard normal distribution.

   (b) Show that the power of the test on the basis of sample size n when $\rho = \rho_1$ where $\rho_1 > 0$ can be calculated by

$$P[Z > z_{\alpha/2} - \sqrt{n-3}\, Z(\rho_1)]. \tag{8.27}$$

14. Using Eq. (8.27), calculate the power of the test for H_0: $\rho = 0$ with a significance level of 5% if $\rho = 0.4$ and $n = 20$.

*15. Consider a one-sided test which limits the probability of rejecting H_0: $\rho = 0$ to α when it is true and not rejecting the H_0 to β when $\rho = \rho_1$ where $\rho_1 > 0$. Show that the required sample size n (in pairs of observations) can be approximated as

$$n = \left(\frac{z_\alpha - z_\beta}{Z(\rho_1)}\right)^2 + 3$$

where z_β is the lower $100\beta\%$ point of the standard normal distribution and $Z(\rho_1)$ is the transformation of ρ_1 given by Eq. (8.18).

16. Referring to the correlation matrix given in Example 8.5, (a) calculate the sample partial correlation coefficient between diastolic pressure and weight, holding the effect of height fixed. Can we conclude that the true partial correlation coefficient is different from zero? (b) What is the correlation between diastolic pressure and height if the effect of weight is removed?

17. Consider the following correlation matrix of yield of hay (X_1), amount of spring rainfall (X_2), and temperature (X_3) for an English area during

20 years. (From 1907 *Journal of the Royal Statistical Society*, 70, p. 1).

$$\begin{array}{c} \\ X_1 \\ X_2 \\ X_3 \end{array} \begin{array}{ccc} X_1 & X_2 & X_3 \\ \left[\begin{array}{ccc} 1 & 0.80 & -0.40 \\ & 1 & -0.56 \\ & & 1 \end{array}\right] \end{array}$$

What interpretation is to be given to an apparent negative correlation between yield and temperature? Answer this question by the method of partial correlation.

18. A neurologic test was given to 27 meningitis patients at the time of their admission and at the end of a treatment period in a children's hospital. In the following data, $+$ and $-$ represent respectively the presence and absence of any neurologic problem. (The data is provided by Dr. Ralph Feigin of Washington University.)

   (a) Test whether the two neurologic test results are correlated.
   (b) Test whether the treatment period reduces the neurologic disturbance of the patient.

Patient	At the time of admission	After treatment	Patient	At the time of admission	After treatment
1	−	−	14	−	−
2	+	−	15	+	+
3	+	−	16	−	−
4	+	+	17	+	−
5	+	+	18	−	−
6	−	+	19	+	−
7	−	+	20	+	+
8	−	−	21	−	−
9	−	+	22	+	+
10	+	−	23	+	+
11	+	+	24	+	−
12	+	+	25	+	−
13	+	−	26	+	+
			27	−	−

19. A sample of twenty 15-year-old girls was selected from a St. Louis city public school. The race and systolic blood pressures of each was recorded in pairs as follows where B and W stand for black and white, respectively:

W	116	W	120	B	108	W	132
B	124	B	112	W	108	W	130
W	122	W	112	B	118	W	114
W	88	W	122	W	120	B	124
W	108	B	100	B	98	W	124

Does this data provide evidence of a correlation between race and blood pressure?

20. The following data, based on a recent FTC report, give the average tar and nicotine contents in mg per cigarette of 12 randomly selected brands of cigarettes. Use a nonparametric test to determine whether the tar and nicotine levels are correlated or not.

Tar	Nicotine	Tar	Nicotine
11	0.6	21	1.6
13	0.8	19	1.3
20	1.4	12	0.8
12	0.9	11	0.8
16	1.1	17	1.3
14	1.0	18	1.3

21. The following data give the heart rate (beats per minute) and total serum cholesterol level (in mg/100 ml) of 20 individuals selected at random from a healthy group of high school boys in the St. Louis area. Determine whether the two variables are correlated by means of rank correlation methods.

Heart Rate	Cholesterol	Heart Rate	Cholesterol	Heart Rate	Cholesterol
75	230	52	230	96	151
67	115	72	176	81	190
79	176	57	171	70	143
74	200	84	180	57	143
57	143	55	129	66	180
66	179	58	225	86	161
66	190	59	161		

*22. Show that (8.10) and (8.17) are algebraically equivalent.

*23. Show that the ordinary correlation coefficient calculated from the ranks of the observations reduces to (8.25).

*24. Show that if $r = 1$, then $r_s = 1$. (Hint: $r = 1$ if, and only if, $Y_i = a + bX_i$ where $b > 0$ for all data points.)

Analysis of Variance

9.1 Introduction

In many experiments the main objective is to determine the effect of one or more external conditions on some response variable, X, of basic or primary interest. For example, in the study of three possible treatments for a certain disease, it may be important to determine the comparative effects of the three treatments on the response variable which might be the survival time, X. Or in the study of tensile strength of rubber compounds, it may be desired to assess the effects of three curing times and four different mixes on the tensile strength, X. In the first example, there is one external condition, namely, the treatments; whereas, in the second example there are two, namely, the curing times and the different mixes. The external conditions, according to which data are classified, are called *factors*. The number of different classes or categories of a factor to be investigated is called the number of *levels* of the factor. Thus, in the first example, the treatment has three levels, and in the second example, the curing time has three levels and the mixes four levels.

The analysis of variance (ANOVA) is a method of investigating the effect of factors on some response variable. To be precise, one may be interested in determining whether the effect of different levels of a factor is significant, and in finding any relation between two or more factors. The analysis of variance constitutes one of the extremely useful and very frequently used methods.

9.2 Models and Assumptions in Analysis of Variance

At this point it is necessary to discuss the various types of factors that have a bearing on the actual analysis. There are two kinds of factors: one with *fixed* effects and the other with *random* effects. The effect of a factor is called fixed if we are working with systematically chosen levels of the factor and we are concerned only with these chosen levels. In the first example in Sec. 9.1 the treatment is a fixed effect if only those three particular treatments are of concern. On the other hand, a factor is said to have a random effect if the given levels of the factor are assumed to be selected randomly from a large or infinite number of levels. In the second example the object was to investigate the effect of changing the curing time which can conceivably be set at many different levels. However, only three levels were to be included in the experiment for the sake of convenience, although inferences about the effects of many possible levels of the factor are hoped to be made from the results of the experiment. It can be assumed that the three levels of the factor are based on a random selection from a large number of levels so the curing time can be considered to have a random effect. Sometimes it is difficult to decide whether a factor is to be regarded as fixed or random. In this case, it would be helpful to question whether the results of the investigation are to be applied only to those particular levels or to all other levels of the factor as well.

Experiments can be classified into three different types: (a) that in which all the factors have fixed effects is called the *fixed effect model*, or *Model I*, (b) that in which all the factors have random effects is called the *random effect model*, or *Model II*, and (c) that with a mixture of factors some fixed and some random is called a *mixed effect model*, or *Model III*.

In applying the analysis of variance, it is necessary to formulate an appropriate statistical model, and to frame the questions in terms of the model. The general form of the ANOVA model can be roughly expressed as:

Observed value = overall mean

$+ \sum$ (term representing effect of appropriate factor)

$+$ (random residual or error term).

The model states that the effects of the factors are additive. Aside from the model, the basic assumptions required in all analyses of variance can be precisely summarized as follows:

(1) Residual or error terms are normally distributed.
(2) Residual or error terms have the same variance.
(3) Residual or error terms are independent.

In general, lack of independency can be relatively more serious than inequality of variances. Even so, the seriousness depends on the magnitude of departures from the assumptions. Nonnormality of the data is less serious in light of the robustness of the test as mentioned in Sec. 7.7. Also nonnormal data are sometimes amenable to transformations as discussed in Sec. 7.7.1.

9.3 Dot Notation for Representing Means

Before discussing the analysis of variance, we shall define some new notation which will be used throughout the chapter. Consider the data which are arranged in the following p by q matrix form:

$$
\begin{matrix}
X_{11} & X_{12} & X_{13} & \cdots & X_{1q} \\
X_{21} & X_{22} & X_{23} & \cdots & X_{2q} \\
X_{31} & X_{32} & X_{33} & \cdots & X_{3q} \\
\cdots & \cdots & \cdots & \cdots & \cdots \\
\cdots & \cdots & \cdots & \cdots & \cdots \\
X_{p1} & X_{p2} & X_{p3} & \cdots & X_{pq}
\end{matrix}
$$

How can we represent, for example, the mean of the observations in the second column or the mean of the observations in the third row, or the overall mean? We introduce the so-called "dot notation" to do this. To be precise,

$X_{i.}$ = sample mean of the ith row

     $= (X_{i1} + X_{i2} + \cdots + X_{iq})/q$,

$X_{.j}$ = sample mean of the jth column

     $= (X_{1j} + X_{2j} + \cdots + X_{pj})/p$,

$X_{..}$ = overall sample mean or the mean of all observations

$$= \sum_{i=1}^{p} \sum_{j=1}^{q} X_{ij}/pq.$$

The following simple numerical example illustrates the usage of the dot notation:

4	12	5	$X_{1.} = 7$
6	12	9	$X_{2.} = 9$
$X_{.1} = 5$	$X_{.2} = 12$	$X_{.3} = 7$	$X_{..} = 8$

Notice the basic mechanism of the notation. For example, $X_{i.}$ denotes the mean calculated by varying only the second subscript while fixing the first subscript at i. The dot notation extends to cases with more than two subscripts. Thus, for example, $X_{.j.}$ represents the mean calculated by varying the first and third subscripts, while the second subscript is fixed at j. Similarly, $X_{i.k}$ denotes the mean calculated by varying the second subscript alone, with the first and third being fixed at i and k, respectively.

9.4 One-Way Classification

Consider an experiment where only one factor is taken into consideration. Let p denote the number of levels for the factor and suppose that $n_1, n_2, \ldots, n_p$ observations are made at the first, the second, ..., and the pth levels, respectively. The numbers of observations may or may not be the same. The data can be arranged in a one-way classification as shown in Table 9.1. The total sample size will be denoted by N throughout this chapter.

TABLE 9.1 Layout of data in one-way classification

	Level			
	1	2	. . .	p
	X_{11}	X_{21}	. . .	X_{p1}
	X_{12}	X_{22}	. . .	X_{p2}
	X_{13}	X_{23}	. . .	X_{p3}
	.	.	. . .	.
	.	.	. . .	.
	.	.	. . .	.
	X_{1n_1}	X_{2n_2}	. . .	X_{pn_p}
Mean	$X_1.$	$X_2.$	. . .	$X_p.$

In Table 9.1, X_{ij} denotes the jth observation made at the ith level or group. The sample mean for the ith level is denoted by $X_i.$ and the overall mean by $X..$. The model for the one-factor experiment expresses X_{ij} as a combination of the overall mean μ, the effect due to the ith level A_i plus the random error ϵ_{ij} for X_{ij}. To be precise,

$$X_{ij} = \mu + A_i + \epsilon_{ij}. \tag{9.1}$$

The error term ϵ_{ij} is considered a normally and independently distributed random effect whose mean value is zero and whose variance is the same for all levels. We can assume with no loss in generality that $\sum A_i = 0$, since if this sum is not zero, it can be made zero by redefining μ. Note that the mean μ_i of X in the ith level is given by $\mu_i = \mu + A_i$.

For both Model I and Model II, the analysis of variance tests the hypothesis that means of all levels are equal. In Model I, since we are concerned only with p particular group means, the hypothesis can be formulated as

$$H_0: \quad \mu_1 = \mu_2 = \cdots = \mu_p, \tag{9.2}$$

or equivalently,

$$H_0: \quad A_1 = A_2 = \cdots = A_p = 0.$$

In Model II, the appropriate hypothesis is given by stating that the variance between the level effects is zero. That is,

$$H_0: \quad \sigma_A^2 = 0, \tag{9.3}$$

since this hypothesis implies that all the levels from which the given levels are randomly selected have the same mean.

The analysis of variance consists of separating the total variation of observations into terms corresponding to factor A and the residual error term ϵ. The total variation of all observations is given by the total sum of square denoted by SS_t which is

$$SS_t = \sum_{i=1}^{p} \sum_{j=1}^{n_i} (X_{ij} - X_{..})^2.$$

Consider the following identity:

$$X_{ij} - X_{..} = (X_{i.} - X_{..}) + (X_{ij} - X_{i.}). \qquad (9.4)$$

The first term on the right-hand side of (9.4) represents the deviation of the ith level mean from the overall mean, so the term estimates A_i which is the effect of the ith level. Next, $X_{ij} - X_{i.}$ is the deviation of the jth individual observation in the ith level from the ith level mean. Thus, the second term on the right-hand side of (9.4) estimates ϵ_{ij} which denotes the residual error of X_{ij}. It is illuminating to note the relation between the parameters in (9.1) and the terms in (9.4) summarized below:

Parameter	Estimator
μ	$X_{..}$
A_i	$X_{i.} - X_{..}$
ϵ_{ij}	$X_{ij} - X_{i.}$

If we square both sides of (9.4) and sum from $j = 1$ to n_i and $i = 1$ to p, we obtain

$$SS_t = \sum_{i=1}^{p} n_i (X_{i.} - X_{..})^2 + \sum_{i=1}^{p} \sum_{j=1}^{n_i} (X_{ij} - X_{i.})^2$$

$$- 2 \sum_{i=1}^{p} \sum_{j=1}^{n_i} (X_{i.} - X_{..})(X_{ij} - X_{i.}). \qquad (9.5)$$

In the summing process, the first term on the right-hand side of (9.5) does not involve the subscript j. Therefore, if we add $(X_{i.} - X_{..})^2$ from $j = 1$ to n_i and $i = 1$ to p, we get $\sum_{i=1}^{p} n_i (X_{i.} - X_{..})^2$. This term is the sum of squares between level means; the sum of squares denoted by SS_b is due to the variation in factor A. The next sum of squares term on the right-hand side represents the part of the sum of squares which is attributed to the residual error term ϵ; the term is called the within-level sum of squares and is denoted by SS_w. Finally, the last term or the cross-product term of (9.5) can be shown to be equal to zero. Thus, (9.5) can be expressed as follows:

$$SS_t = SS_b + SS_w.$$

This shows that, of the total variation in the X's about their overall mean, some of the variation can be ascribed to the variation between level means of the factor and some to the variation of the residual error.

The sum of squares divided by its degrees of freedom is called the *mean square* (MS). The sample variance s^2 given by (4.6) is an example of the mean square; it is a sum of squares divided by $n - 1$ which is its degrees of freedom. The method of determining the degrees of freedom for a sum of squares is beyond the scope of this text. However, the degrees of freedom for SS_b and SS_w are $p - 1$ and $N - p$, respectively.

The basis of the analysis of variance is Theorem 7C, which states that the ratio of two independent mean squares divided by the respective population variances has an F-distribution. In particular, under the common variance, σ^2, the variable

$$F = \frac{(SS_b/p - 1)/\sigma^2}{(SS_w/N - p)/\sigma^2} \tag{9.6}$$

has an F-distribution with $(p - 1, N - p)$ degrees of freedom. This F-ratio, in terms of MS, can be written as

$$F = \frac{MS_b}{MS_w}, \tag{9.7}$$

where MS_b and MS_w represent the "between MS" and "within MS," respectively. Notice that the MS_w is a general form of the pooled variance. See (7.21).

Now, the rejection region of level α for H_0 of either (9.2) or (9.3) is given by

$$F = \frac{MS_b}{MS_w} > F_\alpha$$

where F_α is the upper $100\alpha\%$ of an F-distribution with $(p - 1, N - p)$ degrees of freedom.

All of the constituents of the analysis of variance can be summarized in a so-called *ANOVA table* which is an almost universally accepted form of reporting the computation results. An ANOVA table for a one-way classification has a form shown in Table 9.2.

TABLE 9.2 ANOVA table for one-way classification

Source of variation	Sum of squares	Degrees of freedom	Mean squares	F
Between levels	SS_b	$p - 1$	$MS_b = SS_b/p - 1$	MS_b/MS_w
Within levels	SS_w	$N - p$	$MS_w = SS_w/N - p$	
Total	SS_t	$N - 1$		

In order to further comprehend the rationale of an ANOVA, it is helpful to consider the expected mean square, or $E(MS)$. It represents the expected value of the MS under the model being used. No assumption is made that

H_0 is true in obtaining the $E(MS)$. The $E(MS)$ for the one-way ANOVA is presented in Table 9.3. The derivations of $E(MS)$ are beyond the scope of this book.

TABLE 9.3 $E(MS)$ for one-way classification ANOVA

$$E(MS)$$

Source of variation	Model I	Model II
Between levels	$\sigma^2 + \dfrac{\sum n_t A_t^2}{p-1}$	$\sigma^2 + \dfrac{N^2 - \sum n_t^2}{N(P-1)}\sigma_A^2$
Within levels	σ^2	σ^2

Table 9.3 shows that $E(MS_w) = \sigma^2$. Next, consider $E(MS_b)$ under Model I. If $H_0: A_1 = A_2 = \cdots = A_p = 0$, then $E(MS_b) = \sigma^2$, but if H_0 is false, $E(MS_b) > \sigma^2$. Only if H_0 is true, as has been stated, does the variable F defined by (9.7) have an F-distribution because the common σ^2 cancels out as in (9.6). But if H_0 is not true, the F variable given by (9.7) no longer has an F-distribution, and the important point is that its value would be expected to be larger than one since $E(MS_b) > \sigma^2$ while $E(MS_w) = \sigma^2$. Thus, it would make sense to reject H_0 if the F value is larger than expected by chance. The identical argument holds true for Model II. The rejection of H_0 would indicate that all or some A_t are not equal or $\sigma_A^2 > 0$, depending on the model. In either case, however, the test criterion is given by the ratio $F = MS_b/MS_w$ with $(p - 1, N - p)$ degrees of freedom.

Example 9.1 In order to investigate the possible periodicity (called circadian rhythms) of susceptibility of mice to infection, four groups of mice were given 10^3 fully virulent pneumococci each, at four different times of day. All the mice died from the infection, and the survival times in days were recorded. The data appeared to have a positive skewed distribution, and hence, the logarithmic transformation (base e) was made on each observation. (See Sec. 7.7.1.) Table 9.4 presents the survival time in logarithmic scale. (The data is extracted from an experiment by Dr. Ralph Feigin of Washington University.)

The analysis of variance was performed to compare the mean survival times of four groups each infected at one of the four different times. The ANOVA table is presented in Table 9.5.

Since $F_{0.05} = 2.85$ at $(3, 34)$ degrees of freedom and $3.31 > 2.85$, we reject the hypothesis of equal survival times at the 5% significance level. The analysis suggests that the susceptibility of mice to pneumococcal infection in terms of survival time depends on the time of the day when infected. ▲

TABLE 9.4 Survival time (in $\log_e$ scale) of infected mice

Time of infection	Survival time						
0800	3.33	3.40	3.43	3.49	3.63	3.76	3.85
	3.97	4.04	4.24				
1600	3.29	3.36	3.40	3.43	3.68	3.82	3.87
	4.20	4.49					
2000	3.33	3.36	3.46	3.52	3.61	3.71	3.87
	4.09	4.18					
0400	3.37	3.53	3.76	4.10	4.13	4.22	4.28
	4.55	4.69	5.00				

TABLE 9.5 ANOVA table for data of Table 9.4

Source of variation	Sum of square	Degrees of freedom	Mean square	F
Between groups	1.552	3	0.517	3.31
Within groups	5.312	34	0.156	
Total	6.864	37		

9.5 Multiple Comparison-LSD Test

If the hypothesis of equal means is rejected by an analysis of variance in a one-way classification, we conclude that there are differences among the means. The analysis, however, indicates very little about the nature of the differences, and it does not tell us which means are different from which other means. The problem of examining the differences in more detail by comparing pairs of means is known as *multiple comparison of means*. Several procedures have been proposed for this problem, and readers should refer to Ref. 3 and 9 for more detail. In this section, we shall describe only one method known as the *least significant difference* (*LSD*) test proposed by Fisher for locating those means which are causing a significance difference in ANOVA.

The LSD test has two stages:

Stage 1. Test the hypothesis of equal means by the ANOVA with significance level α.

(a) If the F value is not significant, decide in favor of the hypothesis, and there is nothing more to do.

(b) If the F value is significant, proceed to Stage 2.

Stage 2. Test possible differences between any two groups by an α-level t-test as follows: suppose, without a loss of generality, that the first and second levels are to be compared. The appropriate test criterion is given by the t-statistic we are so familiar with:

$$t = \frac{X_{1.} - X_{2.}}{s_p \sqrt{\dfrac{1}{n_1} + \dfrac{1}{n_2}}} \tag{9.8}$$

where s_p is the pooled sample standard deviation which is given by

$$s_p = \sqrt{MS_w}.$$

The degrees of freedom for the t-statistic is $N - p$, the same as for MS_w.

The test is to reject $\mu_1 = \mu_2$ if $t < -t_{\alpha/2}$ or if $t > t_{\alpha/2}$. Similarly, all $\dfrac{p!}{2!(p-2)!}$ pairs of means can be tested at Stage 2. It should be remarked that these tests are not independent, and for this reason the ANOVA at Stage 1 is essential. Stage 1 guards against falsely rejecting the null hypothesis when it is true. Suppose the first stage is omitted, and we wish to compare five means by performing the t-test at $\alpha = 0.05$ level on each pair of means. There would be ten t values to calculate, and the chance of rejecting at least one of the ten tests will be much larger than 5% even if all samples are from the same population: it can be shown to be about 0.29. With 15 means (105 comparisons among pairs), the probability of finding at least one significant difference is about 0.83 even if all 15 population means are equal. Therefore, the second stage of the LSD test should be used only if the difference is found to be significant by the ANOVA in the first stage.

Example 9.2 Consider the data analyzed in Example 9.1. Recall that we concluded that the difference between four means is significant. An obvious question is which means are different from which other means. The sample means of the four groups were:

$$X_{1.} = 3.71 \qquad X_{2.} = 3.73 \qquad X_{3.} = 3.68 \qquad X_{4.} = 4.17,$$

and $s_p = \sqrt{0.156} = 0.395$. Consider the difference between the first two means. We calculate from (9.8)

$$t = \frac{3.71 - 3.73}{0.395\sqrt{1/10 + 1/9}} = -0.11,$$

and the difference is not significant. Now compare the second and the fourth groups. The t value is

$$t = \frac{3.73 - 4.17}{0.395\sqrt{1/10 + 1/9}} = -2.42,$$

and since $t_{0.025} = 2.03$, the difference is significant at the $P < 0.05$ level based on 34 degrees of freedom. In fact, the LSD test shows that the significant difference between the mean survival times is attributable solely to the prolonged survival time of the animals infected at 0400 hr. This means that the optimal time for being exposed to pneumonia is sometime early in the morning, at least for mice. ▲

*9.6 Two-Way Classification

In many experiments, there is more than one factor involved, and it would be neither economical nor efficient to investigate the effects of the factors one at a time. Among others, such a procedure provides no information on the possible relation between factors. In this section we shall discuss the analysis of the two-way classification design which considers two factors simultaneously. There are two types of two-way classification. The first is "crossed" classification, and the second is "nested" classification.

Let the two factors A and B have p and q levels, respectively. The classification is said to be *crossed* if each level of A is crossed with every level of the second factor B; that is, there is at least one observation in each cell of pq combinations of levels of A and B. On the other hand, if each level of A is associated with a different subset of levels of B, then the factor B is said to be *nested* within A. Perhaps, an example will clarify the definitions. Suppose that one of three machines manufactured by three different companies is to be purchased. In the trial of the machines if each of four men operate each machine at least once, the design is crossed. Suppose that each man operated each machine twice; the amounts produced can be recorded in the form of Table 9.6 which is a typical layout for the two-way crossed design.

TABLE 9.6 Example of two-way crossed classification

Operators	Machine 1		Machine 2		Machine 3	
1	50	48	68	70	34	32
2	47	51	73	69	35	32
3	45	41	54	61	40	39
4	48	50	59	60	39	44

Now, suppose there are not four operators but 12; for example, four operators representing each of the three companies. Then, observations can be presented in the nested form of Table 9.7.

TABLE 9.7 Example of two-way nested classification

	Machine 1				Machine 2				Machine 3			
Operators	1	2	3	4	1	2	3	4	1	2	3	4
Observations	50	47	45	48	68	73	54	59	34	35	40	39
	48	51	41	50	70	69	61	60	32	32	39	44

Note that the data of Table 9.7 may be presented in the form of Table 9.6 for the sake of convenience. However, it is important to see the basic difference between the two designs. Operator 1 for machine 1 is not the same as either of the other operator 1's in the nested classification.

9.6.1 Crossed Classification with Replication

Let there be n observations in each of pq cells of the two-way table, so that there will be $N = npq$ observations in all. Let X_{ijk} be the kth observation in the ith level of A and the jth level of B. The basic model assumed when $n \geq 2$ is

$$X_{ijk} = \mu + A_i + B_j + I_{ij} + \epsilon_{ijk}, \qquad i = 1, 2, \ldots, p;$$
$$j = 1, 2, \ldots, q; \qquad k = 1, 2, \ldots, n, \qquad (9.9)$$

where μ is the overall mean, A_i is the effect due to the ith level of A, B_j is the effect of the jth level of B and I_{ij} is the interaction effect of A_i and B_j, and ϵ_{ijk} is the variation within a particular cell.

The *interaction effect*, I, measures the lack of additivity of effects of two factors. It represents any variations which may be peculiar to a particular combination of A_i and B_j. In Table 9.6, for example, if any significant difference between machines is independent of operators or vice versa, then interaction is not present, and otherwise, interaction is said to be present. The presence or absence of interaction effects can have an important bearing on the interpretation of the experimental result. To illustrate the point, consider the following simple hypothetical experiment in which a conventional internal combustion engine and a new type of engine are compared in terms of gas mileage. Suppose that the data on the experimental mean mileage per gallon (MPG) at three different speeds with all other factors being held identical are given by the following table.

	Speed			
	40	50	60	Mean
Combustion engine	20	17	14	17.0
New engine	19	17	15	17.0
Mean	19.5	17.0	14.5	

The data would indicate that no difference exists between the two types of engines and that both engines tend to be more efficient at the lower speed: the interaction effect is absent, and the interpretation of the experimental result is rather straightforward.

Now suppose that the result turned out as follows:

	Speed			
	40	50	60	Mean
Combustion engine	20	17	14	17.0
New engine	11	19	21	17.0
Mean	15.5	18.0	17.5	

The result would lead to the conclusion that no difference exists between the two engines in terms of efficiency, but that both speed and interaction effects do exist. The interaction effect exists because the new engine performs best at the higher speed while the combustion engine is better at the slower speed. If the type of engine is not specified, then it can be stated on the average that 50 miles is the optimal speed. This interpretation, however, is somewhat complicated in view of the interaction effect. The interaction effect would indicate that varying differences exist in the MPG between the three speeds but that the difference does depend on the type of engine. Thus, in general, the presence of interaction effects implies that a difference attributable to one factor depends on the particular level of the other factor.

Interaction effects can be investigated in a two-way or higher crossed classifications where the experiment is carried out with replication. Now we return to the problem of dealing with a two-way crossed classification with replication.

The ANOVA for a two-way classification can be utilized to test three hypotheses:

(a) The effect of A is zero. That is,
$H_1: A_1 = A_2 = \cdots = A_p = 0$ for a fixed effect,
$H_1: \sigma_A = 0$ for a random effect.
(b) The effect of B is zero. That is,
$H_2: B_1 = B_2 = \cdots = B_q = 0$ for a fixed effect,
$H_2: \sigma_B = 0$ for a random effect.
(c) The interaction effect is zero. That is,
$H_2: I_{ij} = 0$ for all i and j for a fixed effect I,
$H_3: \sigma_I = 0$ for a random effect I.

Of course, for Model I where A and B both have fixed effects, I is also a fixed effect, and in Model II and III, the interaction has a random effect.

The assumptions required in the analysis of variance should be recalled from Sec. 9.2; the residual error term ϵ_{ijk} must be independent and normally distributed with common variance.

The total variation of observations is given by the total sum of squares denoted as SS_t:

$$SS_t = \sum_{ijk} (X_{ijk} - X...)^2$$

where $\sum\limits_{ijk}$ represents three summation terms over all the observations indexed by i, j, and k, i.e., $\sum\limits_{i=1}^{p} \sum\limits_{j=1}^{q} \sum\limits_{k=1}^{n}$.

Consider the following equation which is similar to (9.4):

$$X_{ijk} - X... = (X_{i..} - X...) + (X_{.j.} - X...) + (X_{ij.} - X_{i..} - X_{.j.} + X...)$$
$$+ (X_{ijk} - X_{ij.}) \tag{9.10}$$

On the right-hand side of (9.10), the first term represents the deviation of the ith level mean of the factor A from the overall mean; it estimates A_i. The second term is the deviation of the jth mean of the factor B from the overall mean and it estimates B_j. The third term which estimates I_{ij} measures the interaction effect of A_i and B_j. The last term on the right-hand side is the deviation of the individual observation X_{ijk} from its cell mean; it clearly estimates ϵ_{ijk}. The parameters in the model (9.9) and their estimators can be summarized as follows:

Parameter	Estimator
μ	$X_{...}$
A_i	$X_{i..} - X_{...}$
B_j	$X_{.j.} - X_{...}$
I_{ij}	$X_{ij.} - X_{i..} - X_{.j.} + X_{...}$
ϵ_{ijk}	$X_{ijk} - X_{ij.}$

The relation between (9.9) and (9.10) should become apparent in light of the parameters and estimators given above.

If we square both sides of (9.10) and add over i, j, and k, all cross product terms become zero, and we have, on the right-hand side, four sum of squares terms representing the variation between levels of A, levels of B, interaction effect, and the pooled variation within each cell. The result can be summarized as

$$SS_t = SS_A + SS_B + SS_I + SS_W.$$

These sums of squares together with degrees of freedom can be used to construct the ANOVA table in the form of Table 9.8. In Table 9.8, for example, $\sum_i$ represents the summation over $i = 1$ to p, i.e., $\sum_{i=1}^{p}$. In the squaring process, SS_A is given by $\sum_{ijk}(X_{i..} - X_{...})^2$. However, there is no subscript j or k within the summation sign; thus, over j we add the same expression q times and k we add n times. Therefore, the SS_A can be expressed as $nq \sum_i (X_{i..} - X_{...})^2$.

The other sum of squares are expressed in a similar manner.

Procedures for testing the three hypotheses H_1, H_2, and H_3, which have been formulated in (a), (b), and (c), are based on the F ratios, and the rationale is the same as that discussed in Sec. 9.4. Also, as in Sec. 9.4, the expected mean square, $E(MS)$, shows the precise way of setting up an F-test. The $E(MS)$ for Table 9.8 is summarized in Table 9.9, which gives the $E(MS_A)$, $E(MS_B)$, $E(MS_I)$ and $E(MS_W)$ under the three models.

For purposes of illustration, consider Model II under which we wish to test the three hypotheses designated as H_1, H_2, and H_3. If the H_1 is true, i.e., when $\sigma_A = 0$, then $E(MS_A)$ is identical to $E(MS_I)$. Therefore, the proper

TABLE 9.8 ANOVA table for two-way crossed classification with replication

Source of variation	Sum of squares	Degrees of freedom	Mean squares
A	$SS_A = nq \sum_i (X_{i..} - X_{...})^2$	$p - 1$	$MS_A = SS_A/(p-1)$
B	$SS_B = np \sum_j (X_{.j.} - X_{...})^2$	$q - 1$	$MS_B = SS_B/(q-1)$
I	$SS_I = n \sum_{ij} (X_{ij.} - X_{i..} - X_{.j.} + X_{...})^2$	$(p-1)(q-1)$	$MS_I = SS_I/(p-1)(q-1)$
Within cells	$SS_W = \sum_{ijk} (X_{ijk} - X_{ij.})^2$	$N - pq$	$MS_W = SS_W/(N-pq)$
Total	$SS_t = \sum_{ijk} (X_{ijk} - X_{...})^2$	$N - 1$	

TABLE 9.9 $E(MS)$ for two-way crossed classification with replication

Source of variation	Model I	Model II	Mixed model*
A	$\dfrac{np}{(p-1)} \sum_i A_i^2 + \sigma^2$	$nq\sigma_A^2 + n\sigma_I^2 + \sigma^2$	$\dfrac{nq}{(p-1)} \sum_i A_i^2 + n\sigma_I^2 + \sigma^2$
B	$\dfrac{np}{(q-1)} \sum_j B_j^2 + \sigma^2$	$np\sigma_B^2 + n\sigma_I^2 + \sigma^2$	$np\sigma_B^2 + \sigma^2$
I	$\dfrac{n}{(p-1)(q-1)} \sum_{ij} I_{ij}^2 + \sigma^2$	$n\sigma_I^2 + \sigma^2$	$n\sigma_I^2 + \sigma^2$
Within cells	σ^2	σ^2	σ^2

*Without loss of generality, A is assumed to be fixed and B random.

F-test for the H_1 is made from the ratio of MS_A to MS_I. The situation is identical when H_2 is true. The correct F-test for the H_2 is based on the ratio MS_B/MS_I. Now, consider the test for the H_3 stating $\sigma_I = 0$. When the H_3 is true, $E(MS_I)$ is identical to $E(MS_W)$. Thus, the correct test for the H_3 under Model II is made by calculating the ratio MS_I/MS_W.

Referring to Table 9.9, we can formulate the proper F-test for each hypothesis depending on the model. The correct F-ratios for testing the three hypotheses are summarized as follows:

Hypothesis	Model I	Model II	Mixed model (A fixed and B random)
H_1 (the effect of A)	$F = MS_A/MS_W$	$F = MS_A/MS_I$	$F = MS_A/MS_I$
H_2 (the effect of B)	$F = MS_B/MS_W$	$F = MS_A/MS_I$	$F = MS_B/MS_W$
H_3 (the interaction)	$F = MS_I/MS_W$	$F = MS_I/MS_W$	$F = MS_I/MS_W$

Note that tests for the main factor effects in the mixed model work conversely to the tests in Models I and II. Decision to reject a hypothesis is made if the computed F-value exceeds a critical point of an F-distribution with the proper degrees of freedom. Thus, for example, under Model I the effect of A will be determined to be significant at level α if $MS_A/MS_W > F_\alpha$ where F_α is the upper $100\alpha\%$ point of an F-distribution with $(p - 1, N - pq)$ degrees of freedom.

Example 9.3 An experiment to compare the in vitro effectiveness of three possible antibiotics on three different microorganisms was conducted. The table below summarizes the number of remaining viable organisms from initial inoculum size of about 5×10^4 organisms after eight hours of incubation with one of the three antibiotics. (The data is a part of an experiment conducted by Dr. C. G. Mayhall, Department of Medicine, Washington University.)

Antibiotics	Organism type		
	3445	WE3395	WE3662
Oxacillin	4.1×10^4	1.4×10^4	9.1×10^3
	4.0×10^4	8.5×10^3	8.9×10^3
Cephalothin	7.7×10^4	1.4×10^4	1.4×10^4
	4.4×10^4	7.3×10^3	3.5×10^3
Gentomicin	2.0×10^2	1.0×10^2	5.0×10
	1.0×10^2	1.0×10^2	5.0×10

The two classifications, antibiotics and organism types, are completely crossed with two replications for each combination. Let X_{ijk} denote the kth replicate observation for the jth organism type and the ith antibiotic. The appropriate model takes the form

$$X_{ijk} = \mu + A_i + O_j + I_{ij} + \epsilon_{ijk}$$

where μ is the overall mean, A_i is the effect due to the ith antibiotic, O_j is the effect due to the jth organism type, I_{ij} is the antibiotic's and organism's interaction effect, and ϵ_{ijk} is the variation within a particular cell. The analysis of variance was performed after each observation was transformed using logarithm base 10. (See Sec. 7.7.1 for the rationale for the transformation of the data.) The analysis of variance table calculated from the transformed data is:

Source of variation	S.S.	D.F.	M.S.	F	
Antibiotics	20.295	2	10.148	281.89	$P < 0.001$
Organism type	1.464	2	0.732	20.33	$P < 0.001$
Interaction	0.220	4	0.055	1.53	N.S.
Within cell	0.320	9	0.036		

Since the purpose of the research was to investigate the effectiveness of the three particular drugs on the three specified organism types, the test based on Model I is appropriate. Each of the three F-values is obtained by the ratio of each mean square to the "within cell" mean square. The first F-ratio is tested at $(2, 9)$ degrees of freedom, the next also at $(2, 9)$, and the last F-value at $(4, 9)$ degrees of freedom.

The conclusion is that (a) the differences in effectiveness between the three antibiotics are highly significant, (b) the effectiveness among the three organism types is significantly different, (c) the interaction term is not significant, which would imply that the difference in the effectiveness of the three drugs does not depend on the type of organisms as far as these three organisms are concerned. It is apparent that Gentomicin is uniformly more effective than the other two antibiotics regardless of the organism type. ▲

Prewritten computer programs are available to perform the computations involved in the analysis of variance (see Chap. 10). In using the computer, however, it is important to check that the F-ratio is correctly formed depending on the model assumed.

9.6.2 Crossed Classification with No Replication

Suppose that only a single observation is made in each cell of a two-way classification with p rows and q columns, i.e., $n = 1$. Let X_{ij} be the observation in the ith level of A and the jth level of B. We don't need the third subscript. It turns out that there is no way of testing for an interaction effect when $n = 1$. To be precise, no tests are possible under Model I if the interaction effect exists, whereas under Model II the test is possible for the two main effects but not for the interaction even if the interaction effect presents. Hence, it is convenient to assume a model which does not include the interaction effect, namely,

$$X_{ij} = \mu + A_i + B_j + \epsilon_{ij}$$

where μ is the overall mean, A_i is the effect due to the ith level of the first factor, B_j is the effect due to the jth level of the second factor, and ϵ_{ij} is the term representing the random variation in the ith and jth cells.

Based on the model, we would like to test for the significance of A and B effects, respectively. The appropriate hypotheses are given by the same H_1 and H_2 formulated in Sec. 9.6.1.

Using the squaring and summing processes based on the equality

$$X_{ij} - X_{..} = (X_{i.} - X_{..}) + (X_{.j} - X_{..}) + (X_{ij} - X_{i.} + X_{.j} + X_{..}),$$

the ANOVA table can be derived as in Table 9.10.

Again, the expected mean square, $E(MS)$, will show us how to set up correct F-ratios. $E(MS)$ for the mean square terms of Table 9.10 are sum-

TABLE 9.10 ANOVA table for two-way crossed classification with no replication

Source of variation	Sum of squares	Degrees of freedom	Mean squares
A	$SS_A = q \sum_i (X_{i.} - X_{..})^2$	$p - 1$	$MS_A = SS_A/(p - 1)$
B	$SS_B = p \sum_j (X_{.j} - X_{..})^2$	$q - 1$	$MS_B = SS_B/(q - 1)$
Residual	$SS_R = \sum_{ij} (X_{ij} - X_{i.}$ $- X_{.j} + X_{..})^2$	$(p - 1)(q - 1)$	$MS_R = SS_R/(p - 1)(q - 1)$
Total	$SS_t = \sum_{ij} (X_{ij} - X_{..})^2$	$N - 1$	

marized in Table 9.11. It shows that $E(MS_R) = \sigma^2$. Consider the test for H_1 regarding the effect of the factor A. Regardless of the model we have $E(MS_A) = \sigma^2$ if H_1 is true so that the proper F-test is made from the ratio of MS_A to MS_R. The situation is identical for testing the H_2 about the effect of B.

TABLE 9.11 $E(MS)$ for two-way crossed classification with no replication

Source of variation	Model I	Model II	Mixed model*
A	$q \sum_i A_i^2/(p - 1) + \sigma^2$	$q\sigma_A^2 + \sigma^2$	$q \sum_i A_i^2/(p - 1) + \sigma^2$
B	$p \sum_j B_j^2/(q - 1) + \sigma^2$	$p\sigma_B^2 + \sigma^2$	$p\sigma_B^2 + \sigma^2$
Residuals	σ^2	σ^2	σ^2

*Without loss of generality, A is assumed to be fixed and B random.

The correct F-ratios for testing the two hypotheses are summarized as follows for all models:

Hypothesis	F-ratio	D.F. for F-ratio
H_1 (the effect of A)	$F = MS_A/MR_R$	$p - 1, (p - 1)(q - 1)$
H_2 (the effect of B)	$F = MS_B/MS_R$	$q - 1, (p - 1)(q - 1)$

Example 9.4 An experiment was designed to determine the effect of four different fertilizers on the yield of four varieties of wheat. The following table shows the yield in bushels from unit size lots for each combination of the fertilizer and wheat.

Wheat

Fertilizer	A	B	C	D
1	8	6	6	4
2	7	7	5	4
3	9	9	8	3
4	12	9	8	8

This is an example of crossed classification with no replication. In this design there is no way of testing for an interaction effect even if such an effect may be significant. The model for the experiment is

$$X_{ij} = \mu + F_i + W_j + \epsilon_{ij}$$

where the F_i represents the fertilizer effect and the W_j the effect of the jth variety of wheat. The ANOVA table is calculated in the following:

Source of variation	S.S.	D.F.	M.S.	F	
Fertilizer	30.6	3	10.2	7.85	$P < 0.01$
Wheat	38.6	3	12.9	9.92	$P < 0.01$
Residual	11.7	9	1.3		

The hypothesis of equal fertilizer effect is rejected, and at the same time the difference in the mean yield between the various wheats is determined to be significant. ▲

9.6.3 Nested Classification

Assume that the factor B is nested within A. The basic model for the ANOVA in this case is

$$X_{ijk} = \mu + A_i + B_{j(i)} + \epsilon_{ijk} \tag{9.11}$$

where $B_{j(i)}$ designates the nested effect of B within the effect of A_i. The assumptions are that the ϵ_{ijk} have a normal distribution with common variance σ^2 and $E(\epsilon_{ijk}) = 0$ for all i, j, and k. The nested model has no interaction present since the effects of levels of B are associated with different levels of A. The hypotheses of interest are the significance of the main factor A and that of subfactor B.

The basic equation to be considered is

$$X_{ijk} - X_{...} = (X_{i..} - X_{...}) + (X_{ij.} - X_{i..}) + (X_{ijk} - X_{ij.}). \tag{9.12}$$

Since B is not an independent main factor, the term $(X_{.j.} - X_{...})$ should not be included in the above equality.

By squaring both sides of Eq. (9.12) and summing over i, j, and k, the ANOVA table can be obtained as in Table 9.12.

TABLE 9.12 ANOVA table for nested classification

Source of variation	Sum of squares	Degrees of freedom	Mean squares
A	$SS_A = nq \sum_i (X_{i..} - X_{...})^2$	$p - 1$	$MS_A = SS_A/(p - 1)$
B (within A)	$SS_B = n \sum_{ij} (X_{ij.} - X_{i..})^2$	$p(q - 1)$	$MS_B = SS_B/p(q - 1)$
Within cells	$SS_W = \sum_{ijk} (X_{ijk} - X_{ij.})^2$	$pq(n - 1)$	$MS_W = SS_W/(N - pq)$
Total	$SS_t = \sum_{ijk} (X_{ijk} - X_{...})^2$	$N - 1$	

To set up the proper test we need to look at Table 9.13 which summarizes the $E(MS)$ for different models.

TABLE 9.13 $E(MS)$ for nested classification

Source of variation	Model I	Model II	Mixed model*
A	$nq \sum_i A_i^2/(p - 1) + \sigma^2$	$nq\sigma_A^2 + n\sigma_B^2 + \sigma^2$	$nq \sum_i A_i^2/(p - 1) + \sigma^2$
B	$n \sum_{ij} B_{j(i)}^2/p(q - 1) + \sigma^2$	$n\sigma_B^2 + \sigma^2$	$n\sigma_B^2 + \sigma^2$
Within cells	σ^2	σ^2	σ^2

*It is assumed that A has fixed effects and B random. If the effects of two factors are reversed, then $E(MS)$ is identical to that of Model I with $\sum_i A_i^2/(p - 1)$ replaced by σ_A^2.

Thus, the valid F-test for each hypothesis can be performed from Table 9.12 as follows.

Hypothesis	Model I and mixed model	Model II
H_1 (the effect of A)	$F = MS_A/MA_W$	$F = MS_A/MS_B$
H_2 (the effect of B within A)	$F = MS_B/MS_W$	$F = MS_B/MS_W$

Example 9.5 An experiment was conducted to test the homogeneity of the carbon content using three randomly selected cast iron castings. Three samples were taken from each casting, and each sample was analyzed in duplicate. The percent carbon content is summarized in the following table.

Casting

Samples from A			Samples from B			Samples from C		
1	2	3	1	2	3	1	2	3
1.9	2.2	1.7	3.1	2.8	2.5	3.6	3.2	3.2
1.6	2.5	2.1	2.9	2.8	2.7	3.4	3.5	3.0

The data provides an example of a nested classification characterized by the fact that the samples 1, 2, and 3 are different from one casting to the other. The appropriate model takes the form

$$X_{ijk} = \mu + C_i + S_{j(i)} + \epsilon_{ijk}$$

where the C_i represents the effect associated with the ith casting and the $S_{j(i)}$ the fluctuation effect of the jth sample obtained from the ith casting.

The analysis of variance table is calculated in the table below. The values of F and the P values are based on the assumption that the effects of both castings and samples are random.

Source of variation	S.S.	D.F.	M.S.	F	
Between castings	5.281	2	2.641	22.2	$P < 0.005$
Between samples (within castings)	0.713	6	0.119	3.4	$P < 0.05$
Within samples	0.313	9	0.035		

The first F-ratio obtained by dividing 2.641 by 0.119 is tested at $(2, 6)$ degrees of freedom, and the second F-value obtained by dividing 0.119 by 0.035 is tested at $(6, 9)$ degrees of freedom. The conclusion is that (a) the variation between the true carbon contents of different samples from a given casting is significant at $P < 0.05$, and that (b) the variation in the true carbon contents between the various castings is highly significant, $P < 0.005$. ▲

*9.7 Miscellaneous Remarks on the Analysis of Variance

9.7.1 Low F-Ratios

In order to simplify the discussion, consider the one-way classification analysis of variance as an example. If we look at the expected mean squares given by Table 9.3, we see that the F-ratio will tend to be greater than 1 if the hypothesis is incorrect. Thus, we have used some percentage point in the upper tail of the F-distribution to give us a criterion for rejecting the hypothesis. What happens if the hypothesis is true? In the case of the one-way analysis, both MS_b and MS_w estimate σ^2 as A_i, and σ_A^2 becomes zero depending on the model when the hypothesis is true. Therefore, intuitively, we would anticipate the F-ratio to be about 1. To be more precise, it can be shown that the ratio will be less than 1 at least 50% of the time if the hypothesis is true, and considerably more frequently when the degrees of freedom in the numerator are less than 3 or 4.

Suppose, however, that the calculated F-ratio is so small that it is significant at the lower tail of the F-distribution at the given degrees of freedom. Then, we must suspect something is wrong with the experiment or the model. We may suspect one or both of the following possible errors, especially the first:

(a) The observations are not chosen in a random manner,

(b) The basic additive model of the analysis does not hold.

For example, if the observations are not taken in a random manner, then any factor which is not controlled could increase one mean square while leaving the other mean square unchanged. Thus, a significantly low F-ratio may result.

Example 9.6 Suppose that in Table 9.5 the calculated F-value happened to be 0.10 instead of 3.31 as given. The lower 5% point of the F-distribution at (3, 34) degrees of freedom is calculated using (7.17):

$$\frac{1}{F_{0.05}(34,3)} = \frac{1}{8.61} = 0.116.$$

Thus, the F-ratio is significant at the lower tail, and the validity of the experiment as to the randomness has to be looked into instead of concluding that the hypothesis cannot be rejected. ▲

9.7.2 Missing Data

No problem arises in a one-way classification in the case when some of the observations are missing or lost. The methods described in Sec. 9.4 and Sec. 9.5 do not require an equal number of observations in the various groups.

In all two-way classifications, however, we have supposed the number of observations in each cell to be equal. If this is not true, it is difficult or impossible to separate the total sum of squares into independent components, and the analysis of variance of Sec. 9.6 is no longer vaild. The appropriate analysis is considerably more complicated. Therefore, in designing any experiment, it is a good idea to make, if at all possible, the number of observations in each cell identical. Even so, it will happen occasionally that a few observations are lost. We shall describe two approximate procedures which take care of some common cases of missing data.

(a) Two-way classification with replications when no cell is completely empty:

Suppose that less than about 20% of data is missing. A good approximate method in this case is to replace each missing observation by the mean of the other observations in the same cell. Then, the analysis is carried out as if no data are missing. The only difference is that the degrees of freedom for the "within cell" term is reduced by the number of inserts. To be precise, the degrees of freedom for the within cell term become, for the crossed classification,

$$\text{D.F.} = N - pq - (\text{number of inserts}),$$

and for the nested classification,

$$\text{D.F.} = pq(n - 1) - (\text{number of inserts}),$$

respectively, where N, p, q, and n are as defined in Sec. 9.6.

(b) Two-way crossed classification with no replication when a single observation is missing:

Suppose that X_{ij} is missing. In this case, the method consists of replacing the missing observation by the estimate

$$\hat{X}_{ij} = \frac{pT_i + qT_j - T}{(p - 1)(q - 1)}$$

where T_i denotes the total of $q - 1$ observations in the ith row, T_j, the total of $p - 1$ observations in the jth column, and T, the sum of all $pq - 1$ observations. In applying the method of Sec. 9.6.2, the only difference is that the degrees of freedom for the residual sum of squares are reduced by 1, i.e.,

$$\text{D.F.} = (p - 1)(q - 1) - 1.$$

The above method of estimating a single observation is illustrated next.

Example 9.7 Let us reconsider the data of Example 9.4, where it is now assumed that a single observation is missing as in Table 9.14.

TABLE 9.14 Two-way classification with no replication when single observation is missing

Fertilizer	Wheat			
	A	B	C	D
1	8	6	6	4
2	7	7	5	4
3	9		8	3
4	12	9	8	8

The missing observation X_{32} is estimated by

$$\hat{X}_{32} = \frac{4(20) + 4(22) - 104}{(4 - 1)(4 - 1)} = 7.11.$$

After inserting 7.11 into the missing cell, the analysis can be carried out as in Sec. 9.6.2. The analysis of variance table is given in Table 9.15, where it is noted that the degrees of freedom for the residual term is taken to be 8 instead of 9.

TABLE 9.15 ANOVA table for data of Table 9.14

Source of variation	S.S.	D.F.	M.S.	F	
Fertilizer	30.65	3	10.22	8.59	$P < 0.01$
Wheat	36.76	3	12.25	10.29	$P < 0.01$
Residual	9.55	8	1.19		

Consistent with the finding based on the analysis performed in Example 9.4, we find that both the different fertilizer effect and the difference due to various wheats to be significant at $P < 0.01$. ▲

In using the described methods dealing with missing data, care should be taken to ensure that the degrees of freedom and the within cell or residual mean square term is correct when we use a computer to perform the analysis. We have to readjust the degrees of freedom and recompute the mean square term if the missing data are manually inserted and the computer is used just for the computation of the analysis of variance without properly reducing the degrees of freedom.

It is noted that the above two simple treatments do not exhaust all possible situations with missing or unequal numbers of observations. For example, in a two-way classification with replications, if some of the cells are completely empty, neither of the above methods is valid. See Bennett et al. (Ref. 2) for a more complete and sophisticated method. All troubles and difficulties can be avoided if it is remembered that the analysis of variance does not take kindly to missing or unequal numbers of observations in two or higher classifications.

9.7.3 Greater Numbers of Classifications

The methods of the previous sections can be extended to experiments in which more than two factors are involved. The factors may be either crossed or nested as before. Furthermore, some experiments with three or more factors might involve both crossed and nested classifications. Although the computation becomes more complicated with increasing numbers of factors, the basic rationale and analysis is analogous to that described in Sec. 9.6. However, we shall not discuss the method here because of limited space. Readers should refer to Bennett et al. (Ref. 2) for the analysis of three or more classifications and for more complicated designs.

*9.8 Application of Analysis of Variance to Regression

The principles of the analysis of variance can be applied to test the significance of the regression of a variable Y on the fixed variables, X's. Let the regression line be given by

$$Y = \alpha + \beta_1 X_1 + \beta_2 X_2 + \cdots + \beta_k X_k + \epsilon.$$

It is assumed that Y is normally distributed with common variance about the regression line. The data will consist of n sets of observations which can be displayed in a form of the data matrix:

$$\begin{bmatrix} Y_1 & X_{11} & X_{21} & \cdots & X_{k1} \\ Y_2 & X_{12} & X_{22} & \cdots & X_{k2} \\ \cdot & \cdot & \cdot & \cdots & \cdot \\ \cdot & \cdot & \cdot & \cdots & \cdot \\ \cdot & \cdot & \cdot & \cdots & \cdot \\ Y_n & X_{1n} & X_{2n} & \cdots & X_{kn} \end{bmatrix}$$

Let $\hat{Y}_i$ be the value of Y_i predicted by the estimated regression line $\hat{Y} = a + b_1 X_1 + b_2 X_2 + \cdots + b_k X_k$. We first note that

$$Y_i - \bar{Y} = (\hat{Y}_i - \bar{Y}) + (Y_i - \hat{Y}_i)$$

where $\bar{Y}$ denotes the sample mean calculated from Y_i's. Squaring both sides and summing from $i = 1$ to n, we have

$$\sum (Y_i - \bar{Y})^2 = \sum (\hat{Y}_i - \bar{Y})^2 + \sum (Y_i - \hat{Y}_i)^2 + 2 \sum (\hat{Y}_i - \bar{Y})(Y_i - \hat{Y}_i).$$

The last term can be algebraically shown to be equal to zero. Thus, the above equation reduces to

$$\sum (Y_i - \bar{Y})^2 = \sum (\hat{Y}_i - \bar{Y})^2 + \sum (Y_i - \hat{Y}_i)^2. \qquad (9.13)$$

The term $(Y_i - \bar{Y})$ is the deviation of the ith observation from the overall mean, and so the left-hand side of (9.13) is the sum of squares of the observations from the overall mean or the total sum of squares. Next, $(\hat{Y}_i - \bar{Y})$ is the deviation of the predicted value of the ith observations from the mean. This deviation is due to the regression line, and so the first term in the right-hand side represents the sum of squares due to regression. Finally, the last term represents the sum of squares of the deviations from the regression line, or the part of variation which cannot be accounted for by the regression, and so the term may be called the residual sum of squares. We can summarize the results as follows:

Total sum of squares = sum of squares due to regression

+ residual sum of squares

In other words the total sum of squares can be split into two sums of squares in every regression problem.

Using the technique described in previous sections, we can construct the analysis of variance table in the following form:

TABLE 9.16 Analysis of variance table for regression

Source of variation	Sum of squares	D.F.	Mean square	F
Due to regression	$\sum_i (\hat{Y}_i - \bar{Y})^2$	k	MS_R	MS_R/MS_E
Residual	$\sum_i (Y_i - \hat{Y}_i)^2$	$n - k - 1$	MS_E	
Total	$\sum_i (Y_i - \bar{Y})^2$	$n - 1$		

The last entry, F, can be used to test the significance of a linear regression of Y on X's. To be precise, the hypothesis is $H_0: \beta_1 = \beta_2 = \cdots = \beta_k = 0$. Therefore, the rejection of the hypothesis implies that at least one $\beta_i \neq 0$. Although we will not be able to do it here, it can be shown that the hypothesis $H_0: \beta_1 = \beta_2 = \cdots = \beta_k = 0$ and the hypothesis regarding the population multiple correlation coefficient $H_0: R = 0$ discussed in Sec. 8.7.3 are equivalent. Moreover, the above F-test can be shown to be equivalent to the F-test given in Sec. 8.7.3. In particular, for $k = 1$, the analysis of variance is equivalent to the t-test given by (8.10) or the test formulated in (8.17).

Example 9.8 Table 9.17 below presents the weight (in lb) and fasting and 2-hr plasma glucose values (in mg/100 ml) of ten diabetic patients. Test to determine if the overall regression of the 2-hr glucose value using the weight and fasting value is significant.

TABLE 9.17 Data on weight and glucose values

Weight	Fasting glucose	2-hr glucose
202	99	202
208	294	392
150	438	398
162	249	446
185	107	251
119	86	154
210	120	151
158	124	136
156	133	147
224	106	148

The appropriate model for this problem is assumed to be

$$Y = \alpha + \beta_1 X_1 + \beta_2 X_2 + \epsilon$$

where

Y = 2-hr glucose values,

X_1 = weight,

X_2 = fasting glucose values.

Using a computer program, we obtain

$$\hat{Y} = 39.78 + 0.892X_1 + 0.260X_2,$$

$\hat{R} = 0.83$, $\hat{R}^2 = 0.69$, and $s_{Y \cdot X} = 77.41$. The analysis of variance table is given by

Source of variation	S.S.	D.F.	M.S.	F
Due to regression	93305.9	2	46653.0	7.79
Residual	41946.6	7	5992.4	
Total	135252.5	9		

If we look up percentage points of the F-distribution at $(2,7)$ degrees of freedom, we see that the 5% point is 4.74. Since the calculated value of F exceeds the critical value, that is, $F = 7.79 > 4.74$, we reject the hypothesis H_0: $\beta_1 = \beta_2 = 0$ and conclude that the regression of Y on X_1 and X_2 is significant at $P < 0.05$. We may conclude that the 2-hr glucose value depends on weight and the fasting glucose value and that the latter two variables are valuable in predicting the 2-hr glucose level.

It was stated in Sec. 8.7.2 that the above hypothesis and test are equivalent to a hypothesis $H_0: R = 0$ and the test given by Eq. (8.21). If Eq. (8.21) is used for the above data, we have

$$F = \frac{(7)(0.69)}{2(1 - 0.69)} = 7.79,$$

indicating that the two tests are indeed equivalent. ▲

*9.9 Analysis of Covariance

Before discussing the rationale of the analysis of covariance, we shall consider two examples of situations in which this technique may be appropriate and useful. For the first example, consider that we are concerned with the comparison of two diet programs for losing weight. Suppose that after a fixed period of time one group of women with initial average weight of 200 lb lost an average of 40 lb each while the second group, with a starting mean weight of 160 lb, lost an average of 20 lb each. Then, we do not feel justified in making a direct comparison between 40 lb and 20 lb because the reduction in weight is very likely to depend on the initial weight. Thus, it is necessary to make adjustments on the initial weight difference before we can compare the two diet programs in regard to the weight loss.

As the second example, suppose we are dealing with an experiment performed to compare several methods of reading instruction given to children. The likelihood is that the reading performance of children depends on their initial IQ's. Thus, the difference in IQ's must be compensated for before a fair comparison of reading performance can be made.

The method by which the adjustment and analysis is carried out is known as the analysis of covariance. Suppose it is desired to perform an analysis of variance on a certain variable Y. If Y is related to some secondary variable X, then the dependence of Y on X could obscure and possibly invalidate the results of the analysis of variance performed on Y unless its dependence on X is removed or adjusted. It is convenient to reserve the term *covariable* for a secondary variable that can be used to increase precision in analyzing the primary variable of interest. In the first example above, the weight loss is the primary variable, and the initial weight is the covariable. For the second example, the initial IQ of children can be used as the covariable.

A rather extreme hypothetical situation about the possible effect of a covariable X on the main variable Y can be depicted using Figs. 9.1 and 9.2. In each figure, the actual positions of X are different between two groups

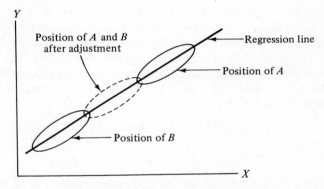

Fig. 9.1 No difference between $\bar{Y}_A$ and $\bar{Y}_B$ after adjustment on X

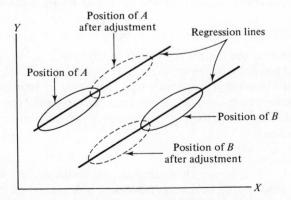

Fig. 9.2 $\bar{Y}_A$ is different from $\bar{Y}_B$ after adjustment on X

A and B. Suppose we graphically move their actual positions parallel to the regression line so their means $\bar{X}$ coincide. In Fig. 9.1, the difference between the mean values of Y, namely, $\bar{Y}_A$ and $\bar{Y}_B$, for the two groups A and B is large, but the difference can be entirely attributed to the difference of $\bar{X}$. If the value of Y is adjusted so that the corresponding values of $\bar{X}$ are the same, then the difference, $\bar{Y}_A - \bar{Y}_B$, is no longer present. In Fig. 9.2, the difference between the two groups is not apparent, but the real difference emerges if the values of Y are adjusted so that the corresponding values of $\bar{X}$ are the same.

The analysis of covariance combines the methods of analysis of variance and regression. This dual role can be illustrated by a one-way classification. We shall assume that the regression line of Y on X for different levels of Y has a common slope β, and that the variance of Y about the regression lines is the same at each level. The relation can be expressed in a regression line as

$$Y_{ij} = \alpha_i + \beta(X_{ij} - X_{..})\epsilon_{ij}.$$

The term $\beta(X_{ij} - X_{..})$ represents the regression effect of Y on X. Thus, $adj. Y_{ij} = Y_{ij} - \beta(X_{ij} - X_{..})$ is the value of Y_{ij} after adjustment by removing the regression effect. Consequently, the analysis of variance performed on $adj. Y_{ij}$ will be independent of the extraneous regression effect. The technique enables us to remove that part of an observed treatment effect which can be attributed to a linear association with X. The analysis of variance model based on the adjusted value of Y is

$$Y_{ij} - \beta(X_{ij} - X_{..}) = \mu + A_i + \epsilon_{ij}$$

or

$$Y_{ij} = \mu + A_i + \beta(X_{ij} - X_{..}) + \epsilon_{ij}.$$

Thus, the analysis of covariance can be regarded as the analysis of variance on Y_{ij} after the linear relation to the covariable is removed. The analysis removes the effect of a source of variation that could otherwise inflate the experimental error. The calculation of the adjusted ith group mean is easily carried out using

$$adj. Y_{i.} = Y_{i.} - b(X_{i.} - X_{..}) \tag{9.14}$$

where b represents the sample regression coefficient.

We shall not describe the computation involved in the analysis of covariance not only because it is complicated, but also, in practice, the actual analysis is performed using a prepared computer program. However, we shall give a formula that can be used to compare only two population means based on the hypothesis

$$H_0: \mu_1 = \mu_2.$$

Let Y_{ij}, $i = 1, 2$, be the jth observation on the ith group for the primary variable and similarly X_{ij} be the corresponding observation on the covariable X. Further, let us use the following notations.

Group	Sample size	Sample mean Y	X	Sample S.D. Y	X	Sample correlation coefficient
1	n_1	$Y_1.$	$X_1.$	s_{1Y}	s_{1X}	r_1
2	n_2	$Y_2.$	$X_2.$	s_{2Y}	s_{2X}	r_2

Note that r_1, for example, denotes the correlation coefficient between X and Y for Group 1 based on n_1 observations. First, the common slope b can be calculated from either

$$b = \frac{\sum_{j=1}^{n_1}(X_{1j} - X_1.)Y_{1j} + \sum_{j=1}^{n_2}(X_{2j} - X_2.)Y_{2j}}{\sum_{j=1}^{n_1}(X_{1j} - X_1.)^2 + \sum_{j=1}^{n_2}(X_{2j} - X_2.)^2} \tag{9.15}$$

or

$$b = \frac{(n_1 - 1)r_1 s_{1X}s_{1Y} + (n_2 - 1)r_2 s_{2X}s_{2Y}}{(n_1 - 1)s_{1X}^2 + (n_2 - 1)s_{2X}^2}.$$

Next, we calculate the adjusted pooled standard deviation by the formula

$$adj.s_p =$$
$$\sqrt{\frac{\sum_{j=1}^{n_1}[(Y_{1j} - Y_1.) - b(X_{1j} - X_1.)]^2 + \sum_{j=1}^{n_2}[(Y_{2j} - Y_2.) - b(X_{2j} - X_2.)]^2}{n_1 + n_2 - 3}} \tag{9.16}$$

or using the equivalent formula given by

$$adj.s_p =$$
$$\sqrt{\frac{(n_1 - 1)(s_{1Y}^2 + b^2 s_{1X}^2 - 2br_1 s_{1X}s_{1Y}) + (n_2 - 1)(s_{2Y}^2 + b^2 s_{2X}^2 - 2br_2 s_{2X}s_{2Y})}{n_1 + n_2 - 3}}$$

The test based on the analysis of covariance reduces to the following t-test statistic

$$t = \frac{(Y_1. - Y_2.) - b(X_1. - X_2.)}{adj.s_p\sqrt{\dfrac{1}{n_1} + \dfrac{1}{n_2}}} \tag{9.17}$$

which has a t-distribution with $n_1 + n_2 - 3$ degrees of freedom.

Note the similarity of formulas (9.16) and (9.17) to (7.21) and (7.24), respectively. Indeed, the relation between the test given by (9.17) to a more general form of analysis of covariance is exactly analogous to that between the two-sample t-test of Sec. 7.5.3 and the analysis of variance.

Example 9.9 Table 9.18 presents the systolic blood pressure and weight of 24 white and 24 black boys all 16 years old randomly selected from St. Louis area schools. We wish to test whether the difference in blood pressure between races in this age group is significant. The mean systolic blood pressure is calculated as 118.33 mm for the white and 116.63 mm for the black.

TABLE 9.18 Systolic blood pressure (in mm Hg) and weight (in lb) of 16-year-old white and black males

White

BP	Wt.	BP	Wt.	BP	Wt.
112	148	104	106	108	120
128	142	118	186	100	119
108	142	114	144	116	171
128	186	128	151	114	115
152	207	140	169	130	132
112	147	96	118	114	109
108	140	124	144	102	132
130	134	126	138	128	141

Black

BP	Wt.	BP	Wt.	BP	Wt.
124	166	100	119	116	142
95	120	112	151	134	179
112	134	134	175	112	133
118	127	148	165	106	140
116	136	110	144	122	149
116	142	112	123	120	149
108	123	126	122	92	129
114	130	130	138	122	133

Since blood pressure and weight are known to be correlated, as illustrated in Example 8.5, a more precise comparison of blood pressure between races should be carried out using the analysis of covariance with weight as a covariable.

Using (9.15) and (9.16), we calculate

$$b = 0.393, \qquad adj.s_p = 9.96.$$

From (9.17) we have $t = 0.18$, which, with 45 degrees of freedom, is not significant at the 5% level of significance. We conclude that the difference between races in regard to blood pressures of 16-year-old boys is not significant.

The mean weight is calculated to be 143.38 lb and 140.38 lb for the whites and blacks, respectively, and the overall mean weight is 141.88 lb. The weight adjusted sample mean blood pressures are calculated from (9.14):

For the whites

$$adj.Y_{1.} = 118.33 - 0.393(143.38 - 141.88) = 117.74,$$

and for the blacks

$$adj.Y_{2.} = 116.63 - 0.393(140.38 - 141.88) = 117.22.$$

These means compare with the unadjusted means of 118.33 and 116.63, respectively. It is observed that the difference between the adjusted means is negligible and is smaller than that of unadjusted means. In other situations, the opposite might be the case.   ▲

PROBLEMS

1. The following table gives the pressure (in lb) in a torsion spring for several settings of the angle between the legs of the spring in a free position.

	Angle of legs	
71°	75°	79°
84	88	92
85	88	92
85	89	93
86	90	

(a) Complete an analysis of variance for this data and state your conclusion.
(b) Perform the LSD test and state your conclusion.

*2. Three automobiles are used in a simple experiment for the purpose of comparing mileage per gallon of many competing brands of gasoline. Three randomly selected brands A, B, and C were used in the experiment. Perform an analysis of variance and draw any pertinent conclusion.

	Gasoline brand		
Car	A	B	C
1	19.4	21.2	20.1
2	14.8	15.1	14.9
3	11.4	10.7	12.3

*3. An electronic components manufacturing firm uses 30 ovens to test the life (in min) of components. The following data are obtained using two randomly selected temperatures and three randomly chosen ovens. Two components were tested simultaneously.

Temperature	Oven 1	Oven 2	Oven 3
1	237	208	192
	254	189	186
2	179	145	127
	182	146	132

Perform an analysis of variance and draw a pertinent conclusion.

*4. The rate of flow of fuel through three different makes of nozzles was investigated in an experiment. Two different operators chosen from a large group were used to test each nozzle. The experiments were made in duplicate.

	Nozzle type		
Operator	A	B	C
1	96.5	96.5	97.1
	97.3	96.1	96.4
2	97.8	96.4	95.9
	97.2	96.8	97.0

Perform an analysis of variance for the given data and draw a pertinent conclusion.

*5. Four different fertilizers and a possible chemical treatment are available for a wheat farm. The following data on yields of wheat were obtained in a greenhouse experiment:

	Fertilizer			
	A	B	C	D
Chemical	21	14	16	12
	23	15	16	13
No chemical	20	12	17	19
	19	13	15	18

Perform an analysis of variance for the data and draw a relevent conclusion.

*6. The following data were obtained in an experiment designed to determine whether there is a difference in viability of progeny of silver salmon according to time of spawning. The variable given is the logarithm of the number of survivors from each female, at the time of downstream migration (rounded for ease in computation).

	Early spawners	Midseason spawners	Late spawners
1973	2.1, 1.5	2.0, 2.4	1.1, 1.0
1974	1.6, 1.8	3.5, 2.5	3.1, 3.0

What may be concluded from these results?

*7. In an experiment at Minter Creek, seven groups of fish with ten fish in each group were divided into four groups and subjected to four different diets. At the end of one and two weeks, blood counts of the fish were taken. The basic data would appear as follows: each entry represents the mean of ten counts. Note that for the diet C only one entry was made.

Week	A	B	C	D
1	76.8	69.7	70.6	78.2
	98.3	93.5		99.4
2	86.2	95.1	106.8	103.5
	101.0	101.6		109.1

Perform all pertinent analyses.

*8. Perform the appropriate analysis of variance for the data given by Table 9.6 and by Table 9.7, respectively. Compare the results of the two analyses.

*9. The following data were taken from a study of blood protein levels in deer. (From 1962 *Syst. Zool.*, 11, p. 131). The variable X is the mobility of serum protein fraction. The sample size in each of the five groups was 12. Perform an analysis of variance and the LSD test.

Group	$\bar{X}$	s
California Blacktail	2.5	0.17
Mule Deer	2.5	0.17
Sitka	2.8	0.24
Vancouver Island Blacktail	2.9	0.17
Whitetail	2.8	0.24

*10. The following data were adopted from a study of the amount X of cotton (in grams) used for nesting materials in both sexes of two subspecies of deer mice, *Peromyscus maniculatus*. (From 1964 *Evolution*, 18, p. 230). The sample size is $n = 24$ for each of the four combinations. Perform an appropriate analysis.

	P. m. bairdii		*P. m. gracilis*	
	$\bar{X}$	s	$\bar{X}$	s
Male	1.7	0.9	2.9	1.4
Female	2.1	1.0	2.6	1.0

*11. Show that the analysis of variance for one-way classification with two groups (i.e., $p = 2$) is equivalent to the two-sample t-test given by (7.26). (Hint: Show that the F-ratio is equal to the square of the t-statistic in this case.)

*12. Show that the analysis of variance described in Sec. 9.8 is equivalent to the t-test given by (8.10) when $k = 1$.

10

Computer Analysis

10.1 Introduction

High-speed electronic computers have triggered a revolution in analyzing data and solving problems in science. In statistical analysis and research, the effect of the computer is three-fold: (a) computational impracticality and complexity is disregarded, (b) the computation is speeded up, and (c) the computation is accurately performed. For these reasons, we are becoming more and more reliant upon the computer for statistical analysis. In a matter of seconds and minutes, vast amounts of statistical computations can be performed. Computations, avoided because of their impracticality prior to the use of computers, can now be handled as a simple routine. The main purpose of this chapter is to describe how we prepare data for computer analysis and to illustrate some case studies. Before we do this, however, we shall briefly describe the main components of a computer system and computer programs.

10.2 Components of a Computer System and Programs

A basic computer system consists of four components—input unit, central processing unit (CPU), storage unit, and output unit. These components are part of every computer system, small or large, even though physically they may be either combined into a single unit or separated into several distinct elements. Fig. 10.1 shows the four components and the flow of information which will be discussed.

The main function of the input unit is to feed both the data and a set of computational instructions, called the program, into the CPU or the storage

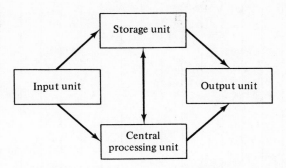

Fig. 10.1   Components of computer system

unit. Common input units are the card reader, the magnetic tape reader, the keyboard, the teletype, among others. The card reader appears to be the most commonly used input device in statistical analysis, although this may change in time.

The function of the storage unit is to store both the data and a sequence of computational instructions. In addition, the storage unit can be used to store intermediate or final results of computations for future use as output. Examples of storage units are magnetic tape, discs, and magnetic cores.

The central processing unit, or CPU, is the main brain of a computer system. It locates both data and instructions and performs actual computations or makes logical decisions.

The fourth component of a computer system is the output device which transmits the output, usually the result of analysis or solutions. The output unit includes printers, teletype, magnetic tape, discs, and many other devices. The most widely used output device in statistical analysis is the printer.

As remarked already, computers are widely used to carry out lengthy mathematical calculations and processing of a large quantity of data. When a computer is used for such purposes, it is necessary to detail for the computer the exact sequence of steps to be followed at each stage in the procedure. Such a sequence of steps designed to perform certain tasks is called a *program*.

Most of the statistical analyses are performed by a set of programs, often referred to as *packaged programs*, already written and stored in a storage unit, usually a disc or tape. Several packaged programs for statistical analyses such as regression analysis, analysis of variance, etc., are maintained in computer centers. Therefore, for most types of statistical analyses we don't have to be concerned about the development of the program. Nevertheless, two major problems confront us in using the program. The first problem is to prepare data in a form that can be fed into a computer. The second is to get access to

TABLE 10.1 Coding sheet

1	2	3	4	5	6	7	8	9	10	11	12	13	14	15	16	17	18	19	20	21	22	23	24	25	...	60	61	62	63	64	65	66	67	68	69	70	71	72	73	74	75	76	77	78	79	80
2	.	4	3			5	6	5					3																																	
3	.	0	5			5	2	5					2																																	
3	.	7	0			5	2	4					2																																	
2	.	5	4			4	9	0					2																																	
3	.	5	3			5	5	5					1																																	
3	.	0	6			5	7	6					2																																	
2	.	8	4			5	5	9					2																																	
1	.	9	0			4	7	9					8																																	
1	.	9	5			5	7	8					4																																	
2	.	2	2			5	4	8					8																																	
3	.	9	0			6	9	6					1																																	
2	.	6	5			6	8	8					5																																	
2	.	3	5			4	6	5					6																																	
2	.	3	3			6	6	0					1																																	
2	.	9	4			4	9	8					1																																	
2	.	2	5			4	7	7					2																																	
2	.	9	8			5	5	2					1																																	
3	.	3	5			5	0	3					1																																	
2	.	5	0			4	6	1					3																																	
1	.	8	5			5	1	7					5																																	

the program which will perform the desired analysis on a given data. The access to the desired program is achieved through certain control cards through the input unit at the time we transmit the data for processing. The method of preparing the control cards for the packaged program is documented in a manual, and we shall not go into a detailed discussion of it. Some of the programs are easy to set up while some require more training and background. In general, however, correct use of the program requires some practice even though knowledge of computer programming may not be needed for most programs.

There are many packaged statistical programs which can be used for standard statistical analyses. A list of some of the most commonly used statistical program packages are listed below.

Program package	Source
BMD	Univ. of California (LA)
SPSS	Univ. of Chicago
SAS	North Carolina State Univ.
STATPACK	Service Bureau Corp.

10.3 Preparing Data for Computer Analysis

When one collects data and records them for computer analysis, it is necessary to code the data in some logical and concise manner which is compatible with the requirements of the computer. Most of the programs in statistical packages require data in a certain form of input format. Here, we shall discuss general principles of preparing data for computer analysis.

The most commonly used form of input is IBM cards, and we shall restrict the discussion to the card form of input. The preparation of data is most conveniently facilitated by a standard coding sheet shown in Table 10.1. Each row of the sheet corresponds to a card, and each column of each row corresponds to a column of a card. The sheet has 80 columns which is the same number as in a card.

In coding the data it is necessary to assign a numerical code to each variable. After this is done, it is essential to determine the maximum anticipated number of digits including a decimal point for each variable. Then, the required number of columns in a coding sheet is assigned to code each variable. To illustrate, consider data consisting of a 4-digit ID number, sex, age in years, weight in pounds, systolic blood pressure, cardiac index, and platelet count per cu mm of blood. The columns can be divided as follows, for example.

Column	Variable	Comment
1–4	ID	
6	sex	$1 =$ male, $2 =$ female
7–8	age	(years)
9–11	weight	(pounds)
13–15	systolic B.P.	
16–19	cardiac index	decimal in column 17
20–21	platelet count	$1 = 0 \leq X < 10^2$
		$2 = 10^2 \leq X < 10^3$
		. . . . .
		. . . . .
		$k = 10^k \leq X < 10^{k+1}$ for any k

Notice that columns 5 and 12 are left blank and one or more blank spaces between variables is optional.

It is absolutely mandatory and essential to follow these general rules:

(1) *Use only those columns assigned for each variable entry.*

(2) *Integers without decimal points must be right justified.* Thus, if an age is 5-years-old, then 5 should be placed in column 8 leaving column 7 blank.

(3) A number with a decimal point can be coded without adhering to rule (2), but it would be more convenient to *keep the decimal point in a prefixed column.*

(4) Missing values can be left as blank or, preferably, may be assigned a special number which cannot possibly occur. So, one may use 9 for sex and 999 for blood pressure, for example.

Sometimes, it is neither convenient nor practical to code the data directly on the standard coding sheet. Such situations occur frequently in survey or biomedical data collection. In that situation it would be convenient to design a special coding protocol tailor-made for the particular project. For example, Table 10.2 shows such a coding form: it was designed for a large high school blood pressure screening project conducted under the auspices of the American Heart Association. Note that the number under the "bar" denotes the column number of a card. Keypunch operators can readily keypunch the data from such a form.

Again, it should be stressed that whether one uses the coding sheet or a special protocol form, it is very essential to adhere to the above four rules. Of course, there is no restriction as to the number of variables. More than one card may be used, but it is often advisable to start each additional card with an ID number and a consecutive card number.

If one of the programs in the program package is to be used in the analy-

TABLE 10.2 Coding form for a blood pressure screening study

HIGH SCHOOL BLOOD PRESSURE SCREENING

NOTE: Place all numbers to the right of the space available.
       Examples: if weight = 94, enter 094.
                if blood pressure = 70/98, enter 070/098.

Code No. ___ ___ ___ ___ ___ ___ ___ ___ ___
         1   3   7
         School #   Student #   Zip Code
                                 (last 3 digits only)

Birthdate ___ ___ ___ ___ ___ ___   Blood Pressure Readings
          10   12   14
          Mo.   Day   Yr.
                                     ___ ___ ___   ___ ___ ___
                                     27   30

Grade   ___ ___
                         16   ___ ___ ___   ___ ___ ___
                                    33   36
Sex   ___
 1. male   18
 2. female   ___ ___ ___   ___ ___ ___
 9. N.A.   39   42
Race   ___
 1. white   19   Time (hr.)   ___ ___
 2. black   45
 3. other
 4. N.A.   1. a.m.
                                     2. p.m.   47

Smoking   ___
 1. yes, < 20/day   20
 2. yes, ≥ 20/day   Date ___ ___ ___ ___ ___ ___
 3. no   48   50   52
 9. N.A.   Mo.   Day   Yr.
                                    Nurse I.D. #
                                                        ___ ___
Height (in.)   ___ ___ ___   54
                   21
                                    Physician care   ___
                                                         56
Weight (lb)   ___ ___ ___
                   24   Varsity sports   ___
                                     1. heavy   57
                                     2. light

---

sis, the user should carefully follow the input specification documented in the
program manual in addition to the general rules discussed with a possible
exception to rule (4). Many packaged programs simply do not allow for any

missing values. To perform a nonstandard analysis, we may have to develop a new program which sometimes is a very time-consuming process.

Finally, it is worth mentioning that the computer does not make all of the decisions. It merely relieves us from tedious and sometimes impossible amounts of computations. The correct interpretation of the results as well as further analysis depends on us. Indeed, we should be cautioned about too strong a dependence on the output of a computer. It is risky to rely blindly on the output, regardless of the assumptions required, circumstances, meanings, models, etc., of the problem. Thus, if a computer program gives all sorts of statistical analyses, this does not necessarily mean that everything is correct or appropriate or that all the information has been extracted from the data.

10.4 Case Studies

In this section, we shall consider three problems, one in regression analysis and the others in the analysis of variance, and the use of a computer for their analyses. The main purpose is to illustrate the typical step-by-step method of interpreting and analyzing the computer printout to make the correct inferences.

Example 10.1 The data recorded in the coding form of Table 10.1 represent the grade-point average (GPA) at the end of the freshman year (in columns 1–5), the college entrance exam score (in columns 6–10) and high school class rank (in columns 11–15) of each of 20 students in a university. Clearly, it would be useful if $Y = $ GPA can be predicted from some combination of $X_1 = $ entrance exam score and $X_2 = $ high school rank. The underlying prediction model is taken to be

$$Y = \alpha + \beta_1 X_1 + \beta_2 X_2 + \epsilon$$

where α, β_1, and β_2 are the coefficients to be calculated and ϵ denotes the random deviation of Y from the regression line.

A program in the BMD computer program package was used to perform the multiple regression analysis. A pertinent part of the computer output is reproduced in Table 10.3.

In the computer printout of Table 10.3, Y, X_1, and X_2 are designated as variables 1, 2, and 3, respectively. The mean and standard deviation of the three variables and the three-by-three correlation matrix are presented. As might be expected, the GPA (Y) is positively correlated with the entrance exam score (X_1), but is negatively with the high school ranking (X_2). The multiple correlation coefficient is $\hat{R} = 0.7747$ from which the coefficient of determination is obtained as 0.60. Thus, about 60% of the total variation in Y is accounted for by the linear association of Y on X_1 and X_2. The standard error of the regression line is seen to be $s_{Y \cdot x} = 0.4026$. The analysis of variance table given next was described in Sec. 9.8. Briefly, it provides an F-ratio for testing the hypothesis that Y is not linearly associated with both X_1 and X_2; that is, $H_0: \beta_1 = 0$ and $\beta_2 = 0$. From Table A4 of the Appendix, the $P = 0.005$ point level of the F-distribution at (2, 17) degrees of freedom is seen to be 7.35. Since the calculated F-ratio is greater than the critical value the hypothesis is rejected at $P < 0.005$. We can conclude that the GPA is linearly associated jointly with the two independent variables. Incidentally, as has been stated in Sec. 8.7.3, this test is equivalent to that of testing the hypothesis of zero multiple correlation coefficient, and this test is given by (8.21). Indeed, the reader can obtain $F = 12.76$ substituting $\hat{R} = 0.7747$ into (8.21).

Finally, from the list of coefficients, the multiple regression equation is obtained as

$$\hat{Y} = 2.024 + 0.0024X_1 - 0.1799X_2.$$

TABLE 10.3 Computer output for Example 10.1

```
PROBLEM CODE                           1
NUMBER OF CASES                        20
NUMBER OF ORIGINAL VARIABLES           3
NUMBER OF VARIABLES ADDED              0
TOTAL NUMBER OF VARIABLES              3
NUMBER OF SUB-PROBLEMS                 1
THE VARIABLE FORMAT IS (4F5.0)


VARIABLE        MEAN              STANDARD DEVIATION
    1          2.76600                  0.60220
    2        544.79980                 69.42506
    3          3.00000                  2.27110


CORRELATION MATRIX

VARIABLE        1           2           3
 NUMBER


    1        1.000       0.394      -0.727
    2                    1.000      -0.181
    3                               1.000

MULTIPLE R                  0.7747
STD. ERROR OF EST.          0.4026

ANALYSIS OF VARIANCE
                    DF   SUM OF SQUARES   MEAN SQUARE   F RATIO
       REGRESSION    2       4.135          2.068       12.760
       RESIDUAL     17       2.755          0.162

                        VARIABLES IN EQUATION

VARIABLE        COEFFICIENT STD. ERROR F TO REMOVE

(CONSTANT       2.02379)
      3         0.00235      0.00135     3.0263 (2)
               -0.17987      0.04134    18.9272 (2)
```

TABLE 10.3 (cont.) Computer output for Example 10.1

LIST OF RESIDUALS

CASE NUMBER	Y X(1)	Y COMPUTED	RESIDUAL	X(3)	X(2)
1	2.4300	2.8135	−0.3835	3.0000	565.0000
2	3.0500	2.8993	0.1507	2.0000	525.0000
3	3.7000	2.8499	0.8501	2.0000	504.0000
4	2.5400	2.8169	−0.2769	2.0000	490.0000
5	3.5300	3.1497	0.3803	1.0000	555.0000
6	3.0600	3.0193	0.0407	2.0000	576.0000
7	2.8400	2.9793	−0.1393	2.0000	559.0000
8	1.9000	1.7118	0.1882	8.0000	479.0000
9	1.9500	2.6642	−0.7142	4.0000	578.0000
10	2.2200	1.8742	0.3458	8.0000	548.0000
11	3.9000	3.4815	0.4185	1.0000	696.0000
12	2.6500	2.7432	−0.0932	5.0000	688.0000
13	2.3500	2.0386	0.3114	6.0000	465.0000
14	3.3300	3.3968	−0.0668	1.0000	660.0000
15	2.9400	3.0156	−0.0756	1.0000	498.0000
16	2.2500	2.7863	−0.5363	2.0000	477.0000
17	2.9800	3.1427	−0.1627	1.0000	552.0000
18	3.3500	3.0274	0.3226	1.0000	503.0000
19	2.5000	2.5688	−0.0688	3.0000	461.0000
20	1.8500	2.3408	−0.4909	5.0000	517.0000

FINISH CARD ENCOUNTERED
PROGRAM TERMINATED

It can be stated that with a given high school rank, 100 points in the exam score is associated with 0.24 point to the GPA on the average, while for fixed exam score each decile rank is associated with −0.18 of the GPA on the average. The final item on the printout is the list of residuals: among other things, it gives Y_i, $\hat{Y}_i$, and $Y_i - \hat{Y}_i$. ▲

In many practical applications, we are confronted with many, say four or more, predictor or independent variables. In such cases, the possibility exists that some of the variables are of little or no value for predictive purposes. Thus, there is a problem of calculating a regression equation which includes only the most pertinent variables. In addition, we should like to limit the number of independent variables because of the cost of monitoring too many variables.

One common approach to this problem is to use the so-called stepwise procedure. The procedure consists of entering independent variables one by one in a stepwise manner on the basis of some criteria on the order of importance of the variables in prediction. Computer programs are available for the stepwise regression analysis.

Example 10.2 The circadian effect of the susceptability of healthy mice to infection was considered in Example 9.1. Further investigation on the subject comparing healthy (sighted) and blind mice was carried out. Six sub-groups of mice from both sighted and blind mice were given 10^3 fully virulent pneumococci at six different times. The survival times in days are summarized in Table 10.4. The following data as well as that of Example 9.1 was extracted from a large set of data from an experiment carried out by Dr. Ralph Feigin of Washington University School of Medicine.

TABLE 10.4 Survival time (in days) of mice infected with pneumococci

Time	Normal mice				Blind mice			
0800	28	33	44	46	35	38	38	39
	49	52	54	55	39	41	43	44
	57	59	62	73	45	49	56	75
1200	31	33	37	45	32	33	34	35
	46	50	52	57	35	38	39	40
	61	69	75	101	43	49	53	91
1600	30	31	31	34	28	29	30	31
	37	43	45	53	32	33	34	39
	57	59	62	67	41	44	52	53
2000	33	40	44	44	30	32	33	33
	48	49	53	53	35	36	37	38
	54	59	61	72	41	42	48	59
2400	32	41	47	47	31	32	32	33
	51	54	55	58	34	35	37	43
	62	70	79	93	52	58	60	65
0400	32	39	45	50	30	31	32	33
	54	57	61	65	33	35	36	38
	73	80	94	105	42	51	56	60

The appropriate analysis of variance model is given by (9.9); it is

$$X_{ijk} = \mu + G_i + T_j + I_{ij} + \epsilon_{ijk}$$

where μ is the overall mean, G_i ($i = 1, 2$) the group effect, T_j ($j = 1, 2, \ldots, 6$) the time effect, I_{ij} the group and time interaction effect, and ϵ_{ijk} the variation within a particular cell. The actual analysis was based on the log transformed data as the survival time exhibited positive skewness. A computer program in the BMD series was used to perform the analysis. A pertinent portion of the computer output is reproduced in Table 10.5.

There are two factors designated as variables in the printout: group (sighted and blind mice) and time (six different times of infection). Thus, the first factor has two levels and the second factor six levels with 12 replicates in each cell. The purpose of the trans-generation cards is to make the desired transformation of observations. In the present example, the trans-generation cards are set up to make log-base e transformation of each observation.

TABLE 10.5 Computer output for Example 10.2

```
PROBLEM NO.  2

NUMBER OF VARIABLES      2
NUMBER OF REPLICATES    12

VARIABLE    NO. OF LEVELS
   1              2
   2              6

     TRANS-GENERATION CARD

CARD NO.    TRANS CODE    CONSTANT
   1            3          0.0
   2            9          2.30260

VARIABLE FORMAT CARD(S)
(3X.F3.0)

GRAND MEAN          3.81293
```

SOURCE OF VARIATION	DEGREES OF FREEDOM	SUMS OF SQUARES	MEAN SQUARES
1	1	5.75595	5.75595
2	5	2.72144	0.54429
12	5	0.20265	0.04053
WITHIN REPLICATES	132	4.23642	0.03209
TOTAL	143	12.91645	

(See Sec. 7.7.1 for the rationale for the transformation.) All computations are performed on the basis of the transformed data. Thus, for example, the grand mean $X_{...} = 3.813$ gives the overall mean of long-transformed observations. The geometric mean, by the way, is calculated to be

$$g_x = \exp(3.813) = 45.28 \text{ (days)}.$$

The most important output in Table 10.5 is, of course, the analysis of variance table corresponding to Table 9.8 calculated next. Clearly, both factors can be considered to have fixed effects, and thus, Model I is appropriate. Using the test given in Sec. 9.6.1, we calculate:

(1) $F = 5.756/0.0321 = 179.3$ to test the difference between the two groups in regard to the average survival time,

(2) $F = 0.544/0.0321 = 16.95$ to test the difference due to the time effect, and

(3) $F = 0.0405/0.0321 = 1.26$ to test the interaction effect.

From the printout, the degrees of freedom for the three F-tests are $(1, 132)$, $(5, 132)$, and $(5, 132)$, respectively. On the basis of the analysis, we conclude that the difference in average survival time between the sighted and blind mice is highly significant; the time of infection has a significant effect, at $P < 0.01$ level. It is rather interesting to find that the interaction effect is not significant, suggesting that the general pattern of the circadian effect of the blind mice is similar to that of the sighted mice even though the average survival time of the blind mice is shorter. ▲

As a final example, we shall illustrate a research problem where the analysis of covariance is applied.

Example 10.3 It has been speculated that the vascular capillary basement membrane thickens in patients with diabetes. In a recent study, the basement membrane width (BMW) in angstrom units was measured on 60 individuals of which half were diabetics. The data consisting of age and BMW of the subjects classified according to sex is presented in Table 10.6. The data is provided by Dr. Joseph Williamson of Washington University.

TABLE 10.6 Age (in Years) and BMW (in Angstrom units) of 60 Individuals

	Normal nondiabetics				Diabetics		
	Male		Female		Male		Female
Age	BMW	Age	BMW	Age	BMW	Age	BMW
71	1129	13	625	22	832	13	701
14	728	65	1013	28	1217	14	642
16	699	22	737	40	2193	50	1125
23	918	71	1329	45	1194	39	667
25	558	24	822	47	837	33	1391
32	910	28	773	49	2114	23	734
35	997	38	966	52	1003	43	825
37	935	39	912	54	1457	48	802
40	1099	43	776	58	1578	17	1301
42	1312	45	924	60	1395	51	1188
45	852	48	700	64	3640	56	960
48	1132	50	662	67	1770	57	1754
50	921	52	1090	78	1853	66	1643
53	859	56	1072	20	705	61	938
59	989	61	941	48	919	64	2118

In addition to comparing the mean BMW of normals and diabetics, it is interesting to see if there is a difference in BMW between the two sexes. The preliminary study indicated that BMW is correlated with age, and therefore, age can be used as a covariable in comparing BMW. The analysis of covariance is an appropriate tool here.

Let Y_{ijk} and X_{ijk} be the kth observation of BMW and age in the ith and jth classification ($i = 1$ for normal, $i = 2$ for diabetics, $j = 1$ for male, and $j = 2$ for female). The appropriate analysis of variance model is

$$Y_{ijk} = \mu + C_i + S_j + I_{ij} + \beta(X_{ijk} - X\dots) + \epsilon_{ijk}$$

where μ is the overall mean, C_i the group effect, S_j the effect due to sexual difference, I_{ij} the interaction effect, $\beta(X_{ijk} - X_{...})$ the regression effect on age, and ϵ_{ijk} the random error. The actual analysis was based on the log transformed BMW values. A pertinent part of the computer output is reproduced in Table 10.7.

TABLE 10.7 Computer output for Example 10.3

```
          PROBLEM 3
          ANALYSIS OF COVARIANCE

   2 VARIABLES  2 FACTORS

      AGE            BMW

   1 CRITERIA    1 COVARIATES WITH THE FOLLOWING VARIABLES
      BMW            AGE

   FACTOR C      2 LEVELS      NORMAL VS DIABETICS
        DEVIATION CONTRASTS

   FACTOR S      2 LEVELS      SEX
        DEVIATION CONTRASTS

                   MEAN AND STANDARD DEVIATION

   FACTOR                                    VARIABLE
    C  S                            AGE        BMW
    1  1     15 OBS.       M       39.333      6.821
                          SD       15.904      0.215

    2  1     15 OBS.       M       48.800      7.225
                          SD       16.341      0.445

    1  2     15 OBS.       M       43.667      6.770
                          SD       16.650      0.209

    2  2     15 OBS.       M       42.333      6.952
                          SD       18.368      0.377

                   ANALYSIS OF VARIANCE TABLE

   SOURCE           SS        DF        MS

   WITHIN CELLS    3.964      55       0.072
   REGRESSION      2.050       1       2.050
   C               0.900       1       0.900
   S               0.337       1       0.337
   CS              0.036       1       0.036
```

Clearly, the problem involves two factors designated as C and S and two variables, age and BMW, where age is used as a covariable. The printout includes the mean and standard deviation of age and of log-transformed BMW classified by the two factors.

The printout given next is the analysis of variance table. From the given MS values, we calculate the following four F-ratios:

(1) $F = 2.050/0.072 = 28.47$ to test the effect of the regression at $(1, 55)$ degrees of freedom;

(2) $F = 0.900/0.072 = 12.50$ to test the difference in the average BMW between normal and diabetic at $(1, 55)$ degrees of freedom;

(3) $F = 0.337/0.072 = 4.68$ to test the difference in the average BMW of male and female; and

(4) $F = 0.036/0.072 = 0.50$ to test the interaction effect.

Based on the four F-tests, we conclude that the regression of BMW on age is significant at $P < 0.005$, so that use of age as a covariable should have improved the precision of analysis. The difference in BMW between normal and diabetic is significant at $P < 0.005$, and that between sexes at $P < 0.05$. According to the table of the means, it is evident that these significant differences have resulted from the larger BMW in diabetics than in normals, and the larger BMW in males than in females. The interaction effect is not significant. Thus, we can state that diabetics have significantly thickened BMW compared to normals and relative size of BMW is larger in males than in females for both normals and diabetics. ▲

References

References Cited in This Text

1. ARMITAGE, P.: *Statistical Methods in Medical Research*, New York: John Wiley & Sons, 1971.

2. BENNETT, C. A., AND FRANKLIN, N. L.: *Statistical Analysis in Chemistry and the Chemical Industry*, New York: John Wiley & Sons, 1954.

3. BROWNLEE, K. A.: *Statistical Theory and Methodology in Science and Engineering*, 2nd ed., New York: John Wiley & Sons, 1965.

4. "Documenta Geigy—Scientific Tables," 7th ed., Geigy Pharmaceuticals, Ardsley, New York, 1973.

5. HOLLANDER, M., AND WOLFE, D. A.: *Nonparametric Statistical Methods*, New York: John Wiley & Sons, 1973.

6. LOEVE, M.: "Fundamental limit theorems of probability theory," *Annals of Math. Stat.* 21, 1950, 321–338.

7. MOOD, A. M., GRAYBILL, F. A., AND BOES, D. C.: *Introduction to the Theory of Statistics*, 3rd ed., New York: McGraw-Hill Book Co., 1974.

8. OSTLE, B.: *Statistics in Research*, 2nd ed., Iowa State Univ. Press, Ames, Iowa, 1963.

9. SNEDECOR, G. W., AND COCHRAN, W. G.: *Statistical Methods*, 6th ed., Iowa State Univ. Press, Ames, Iowa, 1967.

Other Books in Applied Statistics (about the same level as this text)

10. ALDER, H. L., AND ROESSLER, E. B.: *Introduction to Probability and Statistics*, 4th ed., San Francisco: W. H. Freeman and Co., 1970.

11. BOWKER, A. H., AND LIEBERMAN, G. J.: *Engineering Statistics*, Englewood Cliffs, N.J.: Prentice-Hall, Inc., 1959.

12. DIXON, W. J., AND MASSEY, F. J.: *Introduction to Statistical Analysis*, 3rd ed., New York: McGraw-Hill Book Co., 1969.

13. GUENTHER, W. C.: *Concepts of Statistical Inference*, New York: McGraw-Hill Book Co., 1965.

14. GUTTMAN, I., WILKS, S. S., AND HUNTER, J. H.: *Introductory Engineering Statistics*, 2nd ed., New York: John Wiley & Sons, 1971.

15. HODGE, J. L., AND LEHMAN, E. L.: *Basic Concepts of Probability and Statistics*, San Francisco: Holden-Day Inc., 1964.

16. HOEL, P.: *Elementary Statistics*, 3rd ed., New York: John Wiley & Sons, 1971.

17. HUNTSBERGER, D. V.: *Elements of Statistical Inference*, 2nd ed., Boston: Allyn and Bacon Inc., 1967.

18. MODE, E. B.: *Elements of Probability and Statistics*, Englewood Cliffs, N.J.: Prentice-Hall, Inc., 1966.

19. REMINGTON, R. D., AND SCHORK, M. A.: *Statistics with Applications to Biological and Health Sciences*, Englewood Cliffs, N.J.: Prentice-Hall, Inc., 1970.

20. WALPOLE, R. E.: *Introduction to Statistics*, New York: Macmillan Co., 1968.

Other Books in Mathematical Statistics (calculus prerequisite)

21. BRUNK, H. D.: *An Introduction to Mathematical Statistics*, 2nd ed., Waltham, Mass.: Blaisdell Publishing Co., 1965.

22. FREUND, J. E.: *Mathematical Statistics*, 2nd ed., Englewood Cliffs, N.J.: Prentice-Hall, Inc., 1971.

23. HOEL, P.: *Introduction to Mathematical Statistics*, 3rd ed., New York: John Wiley & Sons, 1971.

24. HOEL, P. G., PORT, S. C., AND STONE, C. J.: *Introduction to Statistical Theory*, Boston: Houghton Mifflin Co., 1971.

25. LINDGREN, B. W.: *Statistical Theory*, 2nd ed., New York: Macmillan Co., 1968.

Subscripts and Summations

Mathematical symbols most extensively used in this text are subscripts and the summation notation. For illustration, suppose we arrange the weights of n individuals in a certain order, and let X_1 represent the weight of the first individual, X_2 the weight of the second, and so on, with X_n representing the weight of the nth individual. That is, we can use X_i to denote the weight of ith individual. The index i is called the subscript.

Suppose that we wish to indicate the sum of the weights of the n individuals. It is given by

$$X_1 + X_2 + \cdots + X_n.$$

A more convenient representation for this sum, using the summation symbol $\sum$, is

$$\sum_{i=1}^{n} X_i,$$

and read "summation of X sub i, $i = 1$ to n." Often where there is no cause for confusion as to the number of terms to be added, we shall write more simply

$$\sum X_i$$

omitting the limits of summation. The symbol can be extended to the summation of two or more variables and to more complicated expressions. For example,

$$\sum_{i=1}^{n} (X_i + Y_i) = (X_1 + Y_1) + (X_2 + Y_2) + \cdots + (X_n + Y_n),$$

$$\sum_{i=1}^{n} (X_i - c)^2 = (X_1 + c)^2 + (X_2 + c)^2 + \cdots + (X_n + c)^2$$

where c is a constant.

The following three rules of summation can be easily proved:

Rule 1. The summation of the sum of two or more variables is the sum of their individual summations. Thus,

$$\sum_{i=1}^{n} (X_i + Y_i) = \sum_{i=1}^{n} X_i + \sum_{i=1}^{n} Y_i.$$

Rule 2. A constant factor can be moved across the summation sign. Thus, if c is a constant,

$$\sum_{i=1}^{n} cX_i = c \sum_{i=1}^{n} X_i.$$

Rule 3. The summation of a constant is equal to the product of that constant and the number of terms indicated by the summation sign. Thus,

$$\sum_{i=1}^{n} c = cn.$$

Lists of Tables and Figures

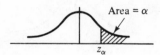

Area = α

z_α

TABLE A1 Area under standard normal distribution
$$\varphi = P\,(Z > z\,\varphi)$$

z_α	0.00	0.01	0.02	0.03	0.04	0.05	0.06	0.07	0.08	0.09
0.0	0.5000	0.4960	0.4920	0.4880	0.4840	0.4801	0.4761	0.4721	0.4681	0.4641
0.1	0.4602	0.4562	0.4522	0.4483	0.4443	0.4404	0.4364	0.4325	0.4286	0.4247
0.2	0.4207	0.4168	0.4129	0.4090	0.4052	0.4013	0.3974	0.3936	0.3897	0.3859
0.3	0.3821	0.3783	0.3745	0.3707	0.3669	0.3632	0.3594	0.3557	0.3520	0.3483
0.4	0.3446	0.3409	0.3372	0.3336	0.3300	0.3264	0.3228	0.3192	0.3156	0.3121
0.5	0.3085	0.3050	0.3015	0.2981	0.2946	0.2912	0.2877	0.2843	0.2810	0.2776
0.6	0.2743	0.2709	0.2676	0.2643	0.2611	0.2578	0.2546	0.2514	0.2483	0.2451
0.7	0.2420	0.2389	0.2358	0.2327	0.2296	0.2266	0.2236	0.2206	0.2177	0.2148
0.8	0.2119	0.2090	0.2061	0.2033	0.2005	0.1977	0.1949	0.1922	0.1894	0.1867
0.9	0.1841	0.1814	0.1788	0.1762	0.1736	0.1711	0.1685	0.1660	0.1635	0.1611
1.0	0.1587	0.1562	0.1539	0.1515	0.1492	0.1469	0.1446	0.1423	0.1401	0.1379
1.1	0.1357	0.1335	0.1314	0.1292	0.1271	0.1251	0.1230	0.1210	0.1190	0.1170
1.2	0.1151	0.1131	0.1112	0.1093	0.1075	0.1056	0.1038	0.1020	0.1003	0.0985
1.3	0.0968	0.0951	0.0934	0.0918	0.0901	0.0885	0.0869	0.0853	0.0838	0.0823
1.4	0.0808	0.0793	0.0778	0.0764	0.0749	0.0735	0.0721	0.0708	0.0694	0.0681
1.5	0.0668	0.0655	0.0643	0.0630	0.0618	0.0606	0.0594	0.0582	0.0571	0.0559
1.6	0.0548	0.0537	0.0526	0.0516	0.0505	0.0495	0.0485	0.0475	0.0465	0.0455
1.7	0.0446	0.0436	0.0427	0.0418	0.0409	0.0401	0.0392	0.0384	0.0375	0.0367
1.8	0.0359	0.0351	0.0344	0.0336	0.0329	0.0322	0.0314	0.0307	0.0301	0.0294
1.9	0.0287	0.0281	0.0274	0.0268	0.0262	0.0256	0.0250	0.0244	0.0239	0.0233
2.0	0.0228	0.0222	0.0217	0.0212	0.0207	0.0202	0.0197	0.0192	0.0188	0.0183
2.1	0.0179	0.0174	0.0170	0.0166	0.0162	0.0158	0.0154	0.0150	0.0146	0.0143
2.2	0.0139	0.0136	0.0132	0.0129	0.0125	0.0122	0.0119	0.0116	0.0113	0.0110
2.3	0.0107	0.0104	0.0102	0.0099	0.0096	0.0094	0.0091	0.0089	0.0087	0.0084
2.4	0.0082	0.0080	0.0078	0.0075	0.0073	0.0071	0.0069	0.0068	0.0066	0.0064
2.5	0.0062	0.0060	0.0059	0.0057	0.0055	0.0054	0.0052	0.0051	0.0049	0.0048
2.6	0.0047	0.0045	0.0044	0.0043	0.0041	0.0040	0.0039	0.0038	0.0037	0.0036
2.7	0.0035	0.0034	0.0033	0.0032	0.0031	0.0030	0.0029	0.0028	0.0027	0.0026
2.8	0.0026	0.0025	0.0024	0.0023	0.0023	0.0022	0.0021	0.0021	0.0020	0.0019
2.9	0.0019	0.0018	0.0018	0.0017	0.0016	0.0016	0.0015	0.0015	0.0014	0.0014
3.0	0.0013	0.0013	0.0013	0.0012	0.0012	0.0011	0.0011	0.0011	0.0010	0.0010
3.1	0.0010	0.0009	0.0009	0.0009	0.0008	0.0008	0.0008	0.0008	0.0007	0.0007
3.2	0.0007	0.0007	0.0006	0.0006	0.0006	0.0006	0.0006	0.0005	0.0005	0.0005
3.3	0.0005	0.0005	0.0005	0.0004	0.0004	0.0004	0.0004	0.0004	0.0004	0.0003
3.4	0.0003	0.0003	0.0003	0.0003	0.0003	0.0003	0.0003	0.0003	0.0003	0.0002
3.5	0.0002	0.0002	0.0002	0.0002	0.0002	0.0002	0.0002	0.0002	0.0002	0.0002
3.6	0.0002	0.0002	0.0001	0.0001	0.0001	0.0001	0.0001	0.0001	0.0001	0.0001
3.7	0.0001	0.0001	0.0001	0.0001	0.0001	0.0001	0.0001	0.0001	0.0001	0.0001
3.8	0.0001	0.0001	0.0001	0.0001	0.0001	0.0001	0.0001	0.0001	0.0001	0.0001
3.9	0.0000	0.0000	0.0000	0.0000	0.0000	0.0000	0.0000	0.0000	0.0000	0.0000
z	0.00	0.01	0.02	0.03	0.04	0.05	0.06	0.07	0.08	0.09

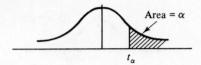

TABLE A2 Percentage point t_α of Student-t distribution

α

D.F.	.40	.30	.20	.10	.05	.025	.010	.005	.001	.0005
1	.325	.727	1.376	3.078	6.314	12.71	31.82	63.66	318.3	636.6
2	.289	.617	1.061	1.886	2.920	4.303	6.965	9.925	22.33	31.60
3	.277	.584	.978	1.638	2.353	3.182	4.541	5.841	10.22	12.94
4	.271	.569	.941	1.533	2.132	2.776	3.747	4.604	7.173	8.610
5	.267	.559	.920	1.476	2.015	2.571	3.365	4.032	5.893	6.859
6	.265	.553	.906	1.440	1.943	2.447	3.143	3.707	5.208	5.959
7	.263	.549	.896	1.415	1.895	2.365	2.998	3.499	4.785	5.405
8	.262	.546	.889	1.397	1.860	2.306	2.896	3.355	4.501	5.041
9	.261	.543	.883	1.383	1.833	2.262	2.821	3.250	4.297	4.781
10	.260	.542	.879	1.372	1.812	2.228	2.764	3.169	4.144	4.587
11	.260	.540	.876	1.363	1.796	2.201	2.718	3.106	4.025	4.437
12	.259	.539	.873	1.356	1.782	2.179	2.681	3.055	3.930	4.318
13	.259	.538	.870	1.350	1.771	2.160	2.650	3.012	3.852	4.221
14	.258	.537	.868	1.345	1.761	2.145	2.624	2.977	3.787	4.140
15	.258	.536	.866	1.341	1.753	2.131	2.602	2.947	3.733	4.073
16	.258	.535	.865	1.337	1.746	2.120	2.583	2.921	3.686	4.015
17	.257	.534	.863	1.333	1.740	2.110	2.567	2.898	3.646	3.965
18	.257	.534	.862	1.330	1.734	2.101	2.552	2.878	3.611	3.922
19	.257	.533	.861	1.328	1.729	2.093	2.539	2.861	3.579	3.883
20	.257	.533	.860	1.325	1.725	2.086	2.528	2.845	3.552	3.850
21	.257	.532	.859	1.323	1.721	2.080	2.518	2.831	3.527	3.819
22	.256	.532	.858	1.321	1.717	2.074	2.508	2.819	3.505	3.792
23	.256	.532	.858	1.319	1.714	2.069	2.500	2.807	3.485	3.767
24	.256	.531	.857	1.318	1.711	2.064	2.492	2.797	3.467	3.745
25	.256	.531	.856	1.316	1.708	2.060	2.485	2.787	3.450	3.725
26	.256	.531	.856	1.315	1.706	2.056	2.479	2.779	3.435	3.707
27	.256	.531	.855	1.314	1.703	2.052	2.473	2.771	3.421	3.690
28	.256	.530	.855	1.313	1.701	2.048	2.467	2.763	3.408	3.674
29	.256	.530	.854	1.311	1.699	2.045	2.462	2.756	3.396	3.659
30	.256	.530	.854	1.310	1.697	2.042	2.457	2.750	3.385	3.646
40	.255	.529	.851	1.303	1.684	2.021	2.423	2.704	3.307	3.551
50	.255	.528	.849	1.298	1.676	2.009	2.403	2.678	3.262	3.495
60	.254	.527	.848	1.296	1.671	2.000	2.390	2.660	3.232	3.460
80	.254	.527	.846	1.292	1.664	1.990	2.374	2.639	3.195	3.415
100	.254	.526	.845	1.290	1.660	1.984	2.365	2.626	3.174	3.389
200	.254	.525	.843	1.286	1.653	1.972	2.345	2.601	3.131	3.339
500	.253	.525	.842	1.283	1.648	1.965	2.334	2.586	3.106	3.310
∞	.253	.524	.842	1.282	1.645	1.960	2.326	2.576	3.090	3.291

Taken from Tables III of Fisher and Yates: *Statistical Tables for Biological, Agricultural, and Medical Research*, published by Longman Group Ltd., London (previously published by Oliver and Boyd, Edinburgh), and by permission of the authors and publishers.

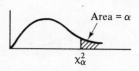

$$\text{Area} = \alpha$$

$$\chi^2_\alpha$$

TABLE A3 Percentage point χ^2_α of chi-square distribution

α

D.F.	.995	.990	.975	.950	.900	.100	.050	.025	.010	.005	.001
1	0.0^{4}393	0.0^{3}157	0.0^{3}982	0.0^{2}393	0.0158	2.71	3.84	5.02	6.63	7.88	10.83
2	0.0100	0.0201	0.0506	0.103	0.211	4.61	5.99	7.38	9.21	10.60	13.82
3	0.072	0.115	0.216	0.352	0.584	6.25	7.81	9.35	11.34	12.84	16.27
4	0.207	0.297	0.484	0.711	1.064	7.78	9.49	11.14	13.28	14.86	18.47
5	0.412	0.554	0.831	1.145	1.61	9.24	11.07	12.83	15.09	16.75	20.52
6	0.676	0.872	1.24	1.64	2.20	10.64	12.59	14.45	16.81	18.55	22.46
7	0.989	1.24	1.69	2.17	2.83	12.02	14.07	16.01	18.48	20.28	24.32
8	1.34	1.65	2.18	2.73	3.49	13.36	15.51	17.53	20.09	21.96	26.13
9	1.73	2.09	2.70	3.33	4.17	14.68	16.92	19.02	21.67	23.59	27.88
10	2.16	2.56	3.25	3.94	4.87	15.99	18.31	20.48	23.21	25.19	29.59
11	2.60	3.05	3.82	4.57	5.58	17.28	19.68	21.92	24.72	26.76	31.26
12	3.07	3.57	4.40	5.23	6.30	18.55	21.03	23.34	26.22	28.30	32.91
13	3.57	4.11	5.01	5.89	7.04	19.81	22.36	24.74	27.69	29.82	34.53
14	4.07	4.66	5.63	6.57	7.79	21.06	23.68	26.12	29.14	31.32	36.12
15	4.60	5.23	6.26	7.26	8.55	22.31	25.00	27.49	30.58	32.80	37.70
16	5.14	5.81	6.91	7.96	9.31	23.54	26.30	28.85	32.00	34.27	39.25
17	5.70	6.41	7.56	8.67	10.09	24.77	27.59	30.19	33.41	35.72	40.79
18	6.26	7.01	8.23	9.39	10.86	25.99	28.87	31.53	34.81	37.16	42.31
19	6.84	7.63	8.91	10.12	11.65	27.20	30.14	32.85	36.19	38.58	43.82
20	7.43	8.26	8.59	10.85	12.44	28.41	31.41	34.17	37.57	40.00	45.32
21	8.03	8.90	10.28	11.59	13.24	29.62	32.67	35.48	38.93	41.40	46.80
22	8.64	9.54	10.98	12.34	14.04	30.81	33.92	36.78	40.29	42.80	48.27
23	9.26	10.20	11.69	13.09	14.85	32.01	35.17	38.08	41.64	44.18	49.73
24	9.89	10.86	12.40	13.85	15.66	33.20	36.42	39.36	42.98	45.56	51.18
25	10.52	11.52	13.12	14.61	16.47	34.38	37.65	40.65	44.31	46.93	52.62
26	11.16	12.20	13.84	15.38	17.29	35.56	38.89	41.92	45.64	48.29	54.05
27	11.81	12.88	14.57	16.15	18.11	36.74	40.11	43.19	46.96	49.64	55.48
28	12.46	13.56	15.31	16.93	18.94	37.92	41.34	44.46	48.28	50.99	56.89
29	13.21	14.26	16.05	17.71	19.77	39.09	42.56	45.72	49.59	52.34	58.30
30	13.79	14.95	16.79	18.49	20.60	40.26	43.77	46.98	50.89	53.67	59.70
40	20.71	22.16	24.43	26.51	29.05	51.80	55.76	59.34	63.69	66.77	73.40
50	27.99	29.71	32.36	34.76	37.69	63.17	67.50	71.42	76.15	79.49	86.66
60	35.53	37.48	40.48	43.19	46.46	74.40	79.08	83.30	88.38	91.95	99.61
70	43.28	45.44	48.76	51.74	55.33	85.53	90.53	95.02	100.4	104.2	112.3
80	51.17	53.54	57.15	60.39	64.28	96.58	101.9	106.6	112.3	116.3	124.8
90	59.20	61.75	65.65	69.13	73.29	107.6	113.1	118.1	124.1	128.3	137.2
100	67.33	70.06	74.22	77.93	82.36	118.5	124.3	129.6	135.8	140.2	149.4

For degrees of freedom $v > 100$, use $\chi^2_\alpha = [(z_\alpha + \sqrt{2v - 1})^2]/2$ where z_α is the corresponding percentage point of the standard normal distribution.

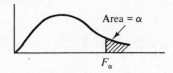

$$\alpha = 0.05$$

n \ m	1	2	3	4	5	6	7	8	9
1	161.45	199.50	215.71	224.58	230.16	233.99	236.77	238.88	240.54
2	18.513	19.000	19.164	19.247	19.296	19.330	19.353	19.371	19.385
3	10.128	9.5521	9.2766	9.1172	9.0135	8.9406	8.8868	8.8452	8.8123
4	7.7086	6.9443	6.5914	6.3883	6.2560	6.1631	6.0942	6.0410	5.9988
5	6.6079	5.7861	5.4095	5.1922	5.0503	4.9503	4.8759	4.8183	4.7725
6	5.9874	5.1433	4.7571	4.5337	4.3874	4.2839	4.2066	4.1468	4.0990
7	5.5914	4.7374	4.3468	4.1203	3.9715	3.8660	3.7870	3.7257	3.6767
8	5.3177	4.4590	4.0662	3.8378	3.6875	3.5806	3.5005	3.4381	3.3881
9	5.1174	4.2565	3.8626	3.6331	3.4817	3.3738	3.2927	3.2296	3.1789
10	4.9646	4.1028	3.7083	3.4780	3.3258	3.2172	3.1355	3.0717	3.0204
11	4.8443	3.9823	3.5874	3.3567	3.2039	3.0946	3.0123	2.9480	2.8962
12	4.7472	3.8853	3.4903	3.2592	3.1059	2.9961	2.9134	2.8486	2.7964
13	4.6672	3.8056	3.4105	3.1791	3.0254	2.9153	2.8321	2.7669	2.7144
14	4.6001	3.7389	3.3439	3.1122	2.9582	2.8477	2.7642	2.6987	2.6458
15	4.5431	3.6823	3.2874	3.0556	2.9013	2.7905	2.7066	2.6408	2.5876
16	4.4940	3.6337	3.2389	3.0069	2.8524	2.7413	2.6572	2.5911	2.5377
17	4.4513	3.5915	3.1968	2.9647	2.8100	2.6987	2.6143	2.5480	2.4943
18	4.4139	3.5546	3.1599	2.9277	2.7729	2.6613	2.5767	2.5102	2.4563
19	4.3808	3.5219	3.1274	2.8951	2.7401	2.6283	2.5435	2.4768	2.4227
20	4.3513	3.4928	3.0984	2.8661	2.7109	2.5990	2.5140	2.4471	2.3928
21	4.3248	3.4668	3.0725	2.8401	2.6848	2.5727	2.4876	2.4205	2.3661
22	4.3009	3.4434	3.0491	2.8167	2.6613	2.5491	2.4638	2.3965	2.3419
23	4.2793	3.4221	3.0280	2.7955	2.6400	2.5277	2.4422	2.3748	2.3201
24	4.2597	3.4028	3.0088	2.7763	2.6207	2.5082	2.4226	2.3551	2.3002
25	4.2417	3.3852	2.9912	2.7587	2.6030	2.4904	2.4047	2.3371	2.2821
26	4.2252	3.3690	2.9751	2.7426	2.5868	2.4741	2.3883	2.3205	2.2655
27	4.2100	3.3541	2.9604	2.7278	2.5719	2.4591	2.3732	2.3053	2.2501
28	4.1960	3.3404	2.9467	2.7141	2.5581	2.4453	2.3593	2.2913	2.2360
29	4.1830	3.3277	2.9340	2.7014	2.5454	2.4324	2.3463	2.2782	2.2229
30	4.1709	3.3158	2.9223	2.6896	2.5336	2.4205	2.3343	2.2662	2.2107
40	4.0848	3.2317	2.8387	2.6060	2.4495	2.3359	2.2490	2.1802	2.1240
60	4.0012	3.1504	2.7581	2.5252	2.3683	2.2540	2.1665	2.0970	2.0401
120	3.9201	3.0718	2.6802	2.4472	2.2900	2.1750	2.0867	2.0164	1.9588
∞	3.8415	2.9957	2.6049	2.3719	2.2141	2.0986	2.0096	1.9384	1.8799

TABLE A4 (cont.) Percentage points F_α of F-distribution

$$\alpha = 0.05$$

n \ m	10	12	15	20	24	30	40	60	120	∞
1	241.88	243.91	245.95	248.01	249.05	250.09	251.14	252.20	253.25	254.32
2	19.396	19.413	19.429	19.446	19.454	19.462	19.471	19.479	19.487	19.496
3	8.7855	8.7446	8.7029	8.6602	8.6385	8.6166	8.5944	8.5720	8.5494	8.5265
4	5.9644	5.9117	5.8578	5.8025	5.7744	5.7459	5.7170	5.6878	5.6581	5.6281
5	4.7351	4.6777	4.6188	4.5581	4.5272	4.4957	4.4638	4.4314	4.3984	4.3650
6	4.0600	3.9999	3.9381	3.8742	3.8415	3.8082	3.7743	3.7398	3.7047	3.6688
7	3.6365	3.5747	3.5108	3.4445	3.4105	3.3758	3.3404	3.3043	3.2674	3.2298
8	3.3472	3.2840	3.2184	3.1503	3.1152	3.0794	3.0428	3.0053	2.9669	2.9276
9	3.1373	3.0729	3.0061	2.9365	2.9005	2.8637	2.8259	2.7872	2.7475	2.7067
10	2.9782	2.9130	2.8450	2.7740	2.7372	2.6996	2.6609	2.6211	2.5801	2.5379
11	2.8536	2.7876	2.7186	2.6464	2.6090	2.5705	2.5309	2.4901	2.4480	2.4045
12	2.7534	2.6866	2.6169	2.5436	2.5055	2.4663	2.4259	2.3842	2.3410	2.2962
13	2.6710	2.6037	2.5331	2.4589	2.4202	2.3803	2.3392	2.2966	2.2524	2.2064
14	2.6021	2.5342	2.4630	2.3879	2.3487	2.3082	2.2664	2.2230	2.1778	2.1307
15	2.5437	2.4753	2.4035	2.3275	2.2878	2.2468	2.2043	2.1601	2.1141	2.0658
16	2.4935	2.4247	2.3522	2.2756	2.2354	2.1938	2.1507	2.1058	2.0589	2.0096
17	2.4499	2.3807	2.3077	2.2304	2.1898	2.1477	2.1040	2.0584	2.0107	1.9604
18	2.4117	2.3421	2.2686	2.1906	2.1497	2.1071	2.0629	2.0166	1.9681	1.9168
19	2.3779	2.3080	2.2341	2.1555	2.1141	2.0712	2.0264	1.9796	1.9302	1.8780
20	2.3479	2.2776	2.2033	2.1242	2.0825	2.0391	1.9938	1.9464	1.8963	1.8432
21	2.3210	2.2504	2.1757	2.0960	2.0540	2.0102	1.9645	1.9165	1.8657	1.8117
22	2.2967	2.2258	2.1508	2.0707	2.0283	1.9842	1.9380	1.8895	1.8380	1.7831
23	2.2747	2.2036	2.1282	2.0476	2.0050	1.9605	1.9139	1.8649	1.8128	1.7570
24	2.2547	2.1834	2.1077	2.0267	1.9838	1.9390	1.8920	1.8424	1.7897	1.7331
25	2.2365	2.1649	2.0889	2.0075	1.9643	1.9192	1.8718	1.8217	1.7684	1.7110
26	2.2197	2.1479	2.0716	1.9898	1.9464	1.9010	1.8533	1.8027	1.7488	1.6906
27	2.2043	2.1323	2.0558	1.9736	1.9299	1.8842	1.8361	1.7851	1.7307	1.6717
28	2.1900	2.1179	2.0411	1.9586	1.9147	1.8687	1.8203	1.7689	1.7138	1.6541
29	2.1768	2.1045	2.0275	1.9446	1.9005	1.8543	1.8055	1.7537	1.6981	1.6377
30	2.1646	2.0921	2.0148	1.9317	1.8874	1.8409	1.7918	1.7396	1.6835	1.6223
40	2.0772	2.0035	1.9245	1.8389	1.7929	1.7444	1.6928	1.6373	1.5766	1.5089
60	1.9926	1.9174	1.8364	1.7480	1.7001	1.6491	1.5943	1.5343	1.4673	1.3893
120	1.9105	1.8337	1.7505	1.6587	1.6084	1.5543	1.4952	1.4290	1.3519	1.2539
∞	1.8307	1.7522	1.6664	1.5705	1.5173	1.4591	1.3940	1.3180	1.2214	1.0000

TABLE A4 (cont.) Percentage points F_α of F-distribution

$$\alpha = 0.025$$

n \ m	1	2	3	4	5	6	7	8	9
1	647.79	799.50	864.16	899.58	921.85	937.11	948.22	956.66	963.28
2	38.506	39.000	39.165	39.248	39.298	39.331	39.355	39.373	39.387
3	17.443	16.044	15.439	15.101	14.885	14.735	14.624	14.540	14.473
4	12.218	10.649	9.9792	9.6045	9.3645	9.1973	9.0741	8.9796	8.9047
5	10.007	8.4336	7.7636	7.3879	7.1464	6.9777	6.8531	6.7572	6.6810
6	8.8131	7.2598	6.5988	6.2272	5.9876	5.8197	5.6955	5.5996	5.5234
7	8.0727	6.5415	5.8898	5.5226	5.2852	5.1186	4.9949	4.8994	4.8232
8	7.5709	6.0595	5.4160	5.0526	4.8173	4.6517	4.5286	4.4332	4.3572
9	7.2093	5.7147	5.0781	4.7181	4.4844	4.3197	4.1971	4.1020	4.0260
10	6.9367	5.4564	4.8256	4.4683	4.2361	4.0721	3.9498	3.8549	3.7790
11	6.7241	5.2559	4.6300	4.2751	4.0440	3.8807	3.7586	3.6638	3.5879
12	6.5538	5.0959	4.4742	4.1212	3.8911	3.7283	3.6065	3.5118	3.4358
13	6.4143	4.9653	4.3472	3.9959	3.7667	3.6043	3.4827	3.3880	3.3120
14	6.2979	4.8567	4.2417	3.8919	3.6634	3.5014	3.3799	3.2853	3.2093
15	6.1995	4.7650	4.1528	3.8043	3.5764	3.4147	3.2934	3.1987	3.1227
16	6.1151	4.6867	4.0768	3.7294	3.5021	3.3406	3.2194	3.1248	3.0488
17	6.0420	4.6189	4.0112	3.6648	3.4379	3.2767	3.1556	3.0610	2.9849
18	5.9781	4.5597	3.9539	3.6083	3.3820	3.2209	3.0999	3.0053	2.9291
19	5.9216	4.5075	3.9034	3.5587	3.3327	3.1718	3.0509	2.9563	2.8800
20	5.8715	4.4613	3.8587	3.5147	3.2891	3.1283	3.0074	2.9128	2.8365
21	5.8266	4.4199	3.8188	3.4754	3.2501	3.0895	2.9686	2.8740	2.7977
22	5.7863	4.3828	3.7829	3.4401	3.2151	3.0546	2.9338	2.8392	2.7628
23	5.7498	4.3492	3.7505	3.4083	3.1835	3.0232	2.9024	2.8077	2.7313
24	5.7167	4.3187	3.7211	3.3794	3.1548	2.9946	2.8738	2.7791	2.7027
25	5.6864	4.2909	3.6943	3.3530	3.1287	2.9685	2.8478	2.7531	2.6766
26	5.6586	4.2655	3.6697	3.3289	3.1048	2.9447	2.8240	2.7293	2.6528
27	5.6331	4.2421	3.6472	3.3067	3.0828	2.9228	2.8021	2.7074	2.6309
28	5.6096	4.2205	3.6264	3.2863	3.0625	2.9027	2.7820	2.6872	2.6106
29	5.5878	4.2006	3.6072	3.2674	3.0438	2.8840	2.7633	2.6686	2.5919
30	5.5675	4.1821	3.5894	3.2499	3.0265	2.8667	2.7460	2.6513	2.5746
40	5.4239	4.0510	3.4633	3.1261	2.9037	2.7444	2.6238	2.5289	2.4519
60	5.2857	3.9253	3.3425	3.0077	2.7863	2.6274	2.5068	2.4117	2.3344
120	5.1524	3.8046	3.2270	2.8943	2.6740	2.5154	2.3948	2.2994	2.2217
∞	5.0239	3.6889	3.1161	2.7858	2.5665	2.4082	2.2875	2.1918	2.1136

TABLE A4 (cont.) Percentage points F_α of F-distribution

$$\alpha = 0.025$$

n \ m	10	12	15	20	24	30	40	60	120	∞
1	968.63	976.71	984.87	993.10	997.25	1001.4	1005.6	1009.8	1014.0	1018.3
2	39.398	39.415	39.431	39.448	39.456	39.465	39.473	39.481	39.490	39.498
3	14.419	14.337	14.253	14.167	14.124	14.081	14.037	13.992	13.947	13.902
4	8.8439	8.7512	8.6565	8.5599	8.5109	8.4613	8.4111	8.3604	8.3092	8.2573
5	6.6192	6.5246	6.4277	6.3285	6.2780	6.2269	6.1751	6.1225	6.0693	6.0153
6	5.4613	5.3662	5.2687	5.1684	5.1172	5.0652	5.0125	5.9589	4.9045	4.8491
7	4.7611	4.6658	4.5678	4.4667	4.4150	4.3624	4.3089	4.2544	4.1989	4.1423
8	4.2951	4.1997	4.1012	3.9995	3.9472	3.8940	3.8398	3.7844	3.7279	3.6702
9	3.9639	3.8682	3.7694	3.6669	3.6142	3.5604	3.5055	3.4493	3.3918	3.3329
10	3.7168	3.6209	3.5217	3.4186	3.3654	3.3110	3.2554	3.1984	3.1399	3.0798
11	3.5257	3.4296	3.3299	3.2261	3.1725	3.1176	3.0613	3.0035	2.9441	2.8828
12	3.3736	3.2773	3.1772	3.0728	3.0187	2.9633	2.9063	2.8478	2.7874	2.7249
13	3.2497	3.1532	3.0527	2.9477	2.8932	2.8373	2.7797	2.7204	2.6590	2.5955
14	3.1469	3.0501	2.9493	2.8437	2.7888	2.7324	2.6742	2.6142	2.5519	2.4872
15	3.0602	2.9633	2.8621	2.7559	2.7706	2.6437	2.5850	2.5242	2.4611	2.3953
16	2.9862	2.8890	2.7875	2.6808	2.6252	2.5678	2.5085	2.4471	2.3831	2.3163
17	2.9222	2.8249	2.7230	2.6158	2.5598	2.5021	2.4422	2.3801	2.3153	2.2474
18	2.8664	2.7689	2.6667	2.5590	2.5027	2.4445	2.3842	2.3214	2.2558	2.1869
19	2.8173	2.7196	2.6171	2.5089	2.4523	2.3937	2.3329	2.2695	2.2032	2.1333
20	2.7737	2.6758	2.5731	2.4645	2.4076	2.3486	2.2873	2.2234	2.1562	2.0853
21	2.7348	2.6368	2.5338	2.4247	2.3675	2.3082	2.2465	2.1819	2.1141	2.0422
22	2.6998	2.6017	2.4984	2.3890	2.3315	2.2718	2.2097	2.1446	2.0760	2.0032
23	2.6682	2.5699	2.4665	2.3567	2.2989	2.2389	2.1763	2.1107	2.0415	1.9677
24	2.6396	2.5412	2.4374	2.3273	2.2693	2.2090	2.1460	2.0799	2.0099	1.9353
25	2.6135	2.5149	2.4110	2.3005	2.2422	2.1816	2.1183	2.0517	1.9811	1.9055
26	2.5895	2.4909	2.3867	2.2759	2.2174	2.1565	2.0928	2.0257	1.9545	1.8781
27	2.5676	2.4688	2.3644	2.2533	2.1946	2.1334	2.0693	2.0018	1.9299	1.8527
28	2.5473	2.4484	2.3438	2.2324	2.1735	2.1121	2.0477	1.9796	1.9072	1.8291
29	2.5286	2.4295	2.3248	2.2131	2.1540	2.0923	2.0276	1.9591	1.8861	1.8072
30	2.5112	2.4120	2.3072	2.1952	2.1359	2.0739	2.0089	1.9400	1.8664	1.7867
40	2.3882	2.2882	2.1819	2.0677	2.0069	1.9429	1.8752	1.8028	1.7242	1.6371
60	2.2702	2.1692	2.0613	1.9445	1.8817	1.8152	1.7440	1.6668	1.5810	1.4822
120	2.1570	2.0548	1.9450	1.8249	1.7597	1.6899	1.6141	1.5299	1.4327	1.3104
∞	2.0483	1.9447	1.8326	1.7085	1.6402	1.5660	1.4835	1.3883	1.2684	1.0000

TABLE A4 (cont.) Percentage points F_α of F-distribution

$$\alpha = 0.01$$

n \ m	1	2	3	4	5	6	7	8	9
1	4052.2	4999.5	5403.3	5624.6	5763.7	5859.0	5928.3	5981.6	6022.5
2	98.503	99.000	99.166	99.249	99.299	99.332	99.356	99.374	99.388
3	34.116	30.817	29.457	28.710	28.237	27.911	27.672	27.489	27.345
4	21.198	18.000	16.694	15.977	15.522	15.207	14.976	14.799	14.659
5	16.258	13.274	12.060	11.392	10.967	10.672	10.456	10.289	10.158
6	13.745	10.925	9.7795	9.1483	8.7459	8.4661	8.2600	8.1016	7.9761
7	12.246	9.5466	8.4513	7.8467	7.4604	7.1914	6.9928	6.8401	6.7188
8	11.259	8.6491	7.5910	7.0060	6.6318	6.3707	6.1776	6.0289	5.9106
9	10.561	8.0215	6.9919	6.4221	6.0569	5.8018	5.6129	5.4671	5.3511
10	10.044	7.5594	6.5523	5.9943	5.6363	5.3858	5.2001	5.0567	4.9424
11	9.6460	7.2057	6.2167	5.6683	5.3160	5.0692	4.8861	4.7445	4.6315
12	9.3302	6.9266	5.9526	5.4119	5.0643	4.8206	4.6395	4.4994	4.3875
13	9.0738	6.7010	5.7394	5.2053	4.8616	4.6204	4.4410	4.3021	4.1911
14	8.8616	6.5149	5.5639	5.0354	4.6950	4.4558	4.2779	4.1399	4.0297
15	8.6831	6.3589	5.4170	4.8932	4.5556	4.3183	4.1415	4.0045	3.8948
16	8.5310	6.2262	5.2922	4.7726	4.4374	4.2016	4.0259	3.8896	3.7804
17	8.3997	6.1121	5.1850	4.6690	4.3359	4.1015	3.9267	3.7910	3.6822
18	8.2854	6.0129	5.0919	4.5790	4.2479	4.0146	3.8406	3.7054	3.5971
19	8.1850	5.9259	5.0103	4.5003	4.1708	3.9386	3.7653	3.6305	3.5225
20	8.0960	5.8489	4.9382	4.4307	4.1027	3.8714	3.6987	3.5644	3.4567
21	8.0166	5.7804	4.8740	4.3688	4.0421	3.8117	3.6396	3.5056	3.3981
22	7.9454	5.7190	4.8166	4.3134	3.9880	3.7583	3.5867	3.4530	3.3458
23	7.8811	5.6637	4.7649	4.2635	3.9392	3.7102	3.5390	3.4057	3.2986
24	7.8229	5.6136	4.7181	4.2184	3.8951	3.6667	3.4959	3.3629	3.2560
25	7.7698	5.5680	4.6755	4.1774	3.8550	3.6272	3.4568	3.3239	3.2172
26	7.7213	5.5263	4.6366	4.1400	3.8183	3.5911	3.4210	3.2884	3.1818
27	7.6767	5.4881	4.6009	4.1056	3.7848	3.5580	3.3882	3.2558	3.1494
28	7.6356	5.4529	4.5681	4.0740	3.7539	3.5276	3.3581	3.2259	3.1195
29	7.5976	5.4205	4.5378	4.0449	3.7254	3.4995	3.3302	3.1982	3.0920
30	7.5625	5.3904	4.5097	4.0179	3.6990	3.4735	3.3045	3.1726	3.0665
40	7.3141	5.1785	4.3126	3.8283	3.5138	3.2910	3.1238	2.9930	2.8876
60	7.0771	4.9774	4.1259	3.6491	3.3389	3.1187	2.9530	2.8233	2.7185
120	6.8510	4.7865	3.9493	3.4796	3.1735	2.9559	2.7918	2.6629	2.5586
∞	6.6349	4.6052	3.7816	3.3192	3.0173	2.8020	2.6393	2.5113	2.4073

TABLE A4 (cont.) Percentage points F_α of F-distribution

$$\alpha = 0.01$$

n \ m	10	12	15	20	24	30	40	60	120	∞
1	6055.8	6106.3	6157.3	6208.7	6234.6	6260.7	6286.8	6313.0	6339.4	6366.0
2	99.399	99.416	99.432	99.449	99.458	99.466	99.474	99.483	99.491	99.501
3	27.229	27.052	26.872	26.690	26.598	26.505	26.411	26.316	26.221	26.125
4	14.546	14.374	14.198	14.020	13.929	13.838	13.745	13.652	13.558	13.463
5	10.051	9.8883	9.7222	9.5527	9.4665	9.3793	9.2912	9.2020	9.1118	9.0204
6	7.8741	7.7183	7.5590	7.3958	7.3127	7.2285	7.1432	7.0568	6.9690	6.8801
7	6.6201	6.4691	6.3143	6.1554	6.0743	5.9921	5.9084	5.8236	5.7372	5.6495
8	5.8143	5.6668	5.5151	5.3591	5.2793	5.1981	5.1156	5.0316	4.9460	4.8588
9	5.2565	5.1114	4.9621	4.8080	4.7290	4.6486	4.5667	4.4831	4.3978	4.3105
10	4.8492	4.7059	4.5582	4.4054	4.3269	4.2469	4.1653	4.0819	3.9965	3.9090
11	4.5393	4.3974	4.2509	4.0990	4.0209	3.9411	3.8596	3.7761	3.6904	3.6025
12	4.2961	4.1553	4.0096	3.8584	3.7805	3.7008	3.6192	3.5355	3.4494	3.3608
13	4.1003	3.9603	3.8154	3.6646	3.5868	3.5070	3.4253	3.3413	3.2548	3.1654
14	3.9394	3.8001	3.6557	3.5052	3.4274	3.3476	3.2656	3.1813	3.0942	3.0040
15	3.8049	3.6662	3.5222	3.3719	3.2940	3.2141	3.1319	3.0471	2.9595	2.8684
16	3.6909	3.5527	3.4089	3.2588	3.1808	3.1007	3.0182	2.9330	2.8447	2.7528
17	3.5931	3.4552	3.3117	3.1615	3.0835	3.0032	2.9205	2.8348	2.7459	2.6530
18	3.5082	3.3706	3.2273	3.0771	2.9990	2.9185	2.8354	2.7493	2.6597	2.5660
19	3.4338	3.2965	3.1533	3.0031	2.9249	2.8442	2.7608	2.6742	2.5839	2.4893
20	3.3682	3.2311	3.0880	2.9377	2.8594	2.7785	2.6947	2.6077	2.5168	2.4212
21	3.3098	3.1729	3.0299	2.8796	2.8011	2.7200	2.6359	2.5484	2.4568	2.3603
22	3.2576	3.1209	2.9780	2.8274	2.7488	2.6675	2.5831	2.4951	2.4029	2.3055
23	3.2106	3.0740	2.9311	2.7805	2.7017	2.6202	2.5355	2.4471	2.3542	2.2559
24	3.1681	3.0316	2.8887	2.7380	2.6591	2.5773	2.4923	2.4035	2.3099	2.2107
25	3.1294	2.9931	2.8502	2.6993	2.6203	2.5383	2.4530	2.3637	2.2695	2.1694
26	3.0941	2.9579	2.8150	2.6640	2.5848	2.5026	2.4170	2.3273	2.2325	2.1315
27	3.0618	2.9256	2.7827	2.6316	2.5522	2.4699	2.3840	2.2938	2.1984	2.0965
28	3.0320	2.8959	2.7530	2.6017	2.5223	2.4397	2.3535	2.2629	2.1670	2.0642
29	3.0045	2.8685	2.7256	2.5742	2.4946	2.4118	2.3253	2.2344	2.1378	2.0342
30	2.9791	2.8431	2.7002	2.5487	2.4689	2.3860	2.2992	2.2079	2.1107	2.0062
40	2.8005	2.6648	2.5216	2.3689	2.2880	2.2034	2.1142	2.0194	1.9172	1.8047
60	2.6318	2.4961	2.3523	2.1978	2.1154	2.0285	1.9360	1.8363	1.7263	1.6006
120	2.4721	2.3363	2.1915	2.0346	1.9500	1.8600	1.7628	1.6557	1.5330	1.3805
∞	2.3209	2.1848	2.0385	1.8783	1.7908	1.6964	1.5923	1.4730	1.3246	1.0000

TABLE A4 (cont.) Percentage points F_α of F-distribution

$$\alpha = 0.005$$

n \ m	1	2	3	4	5	6	7	8	9
1	16211	20000	21615	22500	23056	23437	23715	23925	24091
2	198.50	199.00	199.17	199.25	199.30	199.33	199.36	199.37	199.39
3	55.552	49.799	47.467	46.195	45.392	44.838	44.434	44.126	43.882
4	31.333	26.284	24.259	23.155	22.456	21.975	21.622	21.352	21.139
5	22.785	18.314	16.530	15.556	14.940	14.513	14.200	13.961	13.772
6	18.635	14.544	12.917	12.028	11.464	11.073	10.786	10.566	10.391
7	16.236	12.404	10.882	10.050	9.5221	9.1554	8.8854	8.6781	8.5138
8	14.688	11.042	9.5965	8.8051	8.3018	7.9520	7.6942	7.4960	7.3386
9	13.614	10.107	8.7171	7.9559	7.4711	7.1338	6.8849	6.6933	6.5411
10	12.826	9.4270	8.0807	7.3428	6.8723	6.5446	6.3025	6.1159	5.9676
11	12.226	8.9122	7.6004	6.8809	6.4217	6.1015	5.8648	5.6821	5.5368
12	11.754	8.5096	7.2258	6.5211	6.0711	5.7570	5.5245	5.3451	5.2021
13	11.374	8.1865	6.9257	6.2335	5.7910	5.4819	5.2529	5.0761	4.9351
14	11.060	7.9217	6.6803	5.9984	5.5623	5.2574	5.0313	4.8566	4.7173
15	10.798	7.7008	6.4760	5.8029	5.3721	5.0708	4.8473	4.6743	4.5364
16	10.575	7.5138	6.3034	5.6378	5.2117	4.9134	4.6920	4.5207	4.3838
17	10.384	7.3536	6.1556	5.4967	5.0746	4.7789	4.5594	4.3893	4.2535
18	10.218	7.2148	6.0277	5.3746	4.9560	4.6627	4.4448	4.2759	4.1410
19	10.073	7.0935	5.9161	5.2681	4.8526	4.5614	4.3448	4.1770	4.0428
20	9.9439	6.9865	5.8177	5.1743	4.7616	4.4721	4.2569	4.0900	3.9564
21	9.8295	6.8914	5.7304	5.0911	4.6808	4.3931	4.1789	4.0128	3.8799
22	9.7271	6.8064	5.6524	5.0168	4.6088	4.3225	4.1094	3.9440	3.8116
23	9.6348	6.7300	5.5823	4.9500	4.5441	4.2591	4.0469	3.8822	3.7502
24	9.5513	6.6610	5.5190	4.8898	4.4857	4.2019	3.9905	3.8264	3.6949
25	9.4753	6.5982	5.4615	4.8351	4.4327	4.1500	3.9394	3.7758	3.6447
26	9.4059	6.5409	5.4091	4.7852	4.3844	4.1027	3.8928	3.7297	3.5989
27	9.3423	6.4885	5.3611	4.7396	4.3402	4.0594	3.8501	3.6875	3.5571
28	9.2838	6.4403	5.3170	4.6977	4.2996	4.0197	3.8110	3.6487	3.5186
29	9.2297	6.3958	5.2764	4.6591	4.2622	3.9830	3.7749	3.6130	3.4832
30	9.1797	6.3547	5.2388	4.6233	4.2276	3.9492	3.7416	3.5801	3.4505
40	8.8278	6.0664	4.9759	4.3738	3.9860	3.7129	3.5088	3.3498	3.2220
60	8.4946	5.7950	4.7290	4.1399	3.7600	3.4918	3.2911	3.1344	3.0083
120	8.1790	5.5393	4.4973	3.9207	3.5482	3.2849	3.0874	2.9330	2.8083
∞	7.8794	5.2983	4.2794	3.7151	3.3499	3.0913	2.8968	2.7444	2.6210

TABLE A4 (cont.) Percentage points F_α of F-distribution

$$\alpha = 0.005$$

n \ m	10	12	15	20	24	30	40	60	120	∞
1	24224	24426	24630	24836	24940	25044	25148	25253	25359	25465
2	199.40	199.42	199.43	199.45	199.46	199.47	199.47	199.48	199.49	199.51
3	43.686	43.387	43.085	42.778	42.622	42.466	42.308	42.149	41.989	41.829
4	20.967	20.705	20.438	20.167	20.030	19.892	19.752	19.611	19.468	19.325
5	13.618	13.384	13.146	12.903	12.780	12.656	12.530	12.402	12.274	12.144
6	10.250	10.034	9.8140	9.5888	9.4741	9.3583	9.2408	9.1219	9.0015	8.8793
7	8.3803	8.1764	7.9678	7.7540	7.6450	7.5345	7.4225	7.3088	7.1933	7.0760
8	7.2017	7.0149	6.8143	6.6082	6.5029	6.3961	6.2875	6.1772	6.0649	5.9505
9	6.4171	6.2274	6.0325	5.8318	5.7292	5.6248	5.5186	5.4104	5.3001	5.1875
10	5.8467	5.6613	5.4707	5.2740	5.1732	5.0705	4.9659	4.8592	4.7501	4.6385
11	5.4182	5.2363	5.0489	4.8552	4.7557	4.6543	4.5508	4.4450	4.3367	4.2256
12	5.0855	4.9063	4.7214	4.5299	4.4315	4.3309	4.2282	4.1229	4.0149	3.9039
13	4.8199	4.6429	4.4600	4.2703	4.1726	4.0727	3.9704	3.8655	3.7577	3.6465
14	4.6034	4.4281	4.2468	4.0585	3.9614	3.8619	3.7600	3.6553	3.5473	3.4359
15	4.4236	4.2498	4.0698	3.8826	3.7859	3.6867	3.5850	3.4803	3.3722	3.2602
16	4.2719	4.0994	3.9205	3.7342	3.6378	3.5388	3.4372	3.3324	3.2240	3.1115
17	4.1423	3.9709	3.7929	3.6073	3.5112	3.4124	3.3107	3.2058	3.0971	2.9839
18	4.0305	3.8599	3.6827	3.4977	3.4017	3.3030	3.2014	3.0962	2.9871	2.8732
19	3.9329	3.7631	3.5866	3.4020	3.3062	3.2075	3.1058	3.0004	2.8908	2.7762
20	3.8470	3.6779	3.5020	3.3178	3.2220	3.1234	3.0215	2.9159	2.8058	2.6904
21	3.7709	3.6024	3.4270	3.2431	3.1474	3.0488	2.9467	2.8408	2.7302	2.6140
22	3.7030	3.5350	3.3600	3.1764	3.0807	2.9821	2.8799	2.7736	2.6625	2.5455
23	3.6420	3.4745	3.2999	3.1165	3.0208	2.9221	2.8198	2.7132	2.6016	2.4837
24	3.5870	3.4199	3.2456	3.0624	2.9667	2.8679	2.7654	2.6585	2.5463	2.4276
25	3.5370	3.3704	3.1963	3.0133	2.9176	2.8187	2.7160	2.6088	2.4960	2.3765
26	3.4916	3.3252	3.1515	2.9685	2.8728	2.7738	2.6709	2.5633	2.4501	2.3297
27	3.4499	3.2839	3.1104	2.9275	2.8318	2.7327	2.6296	2.5217	2.4078	2.2867
28	3.4117	3.2460	3.0727	2.8899	2.7941	2.6949	2.5916	2.4834	2.3689	2.2469
29	3.3765	3.2111	3.0379	2.8551	2.7594	2.6601	2.5565	2.4479	2.3330	2.2102
30	3.3440	3.1787	3.0057	2.8230	2.7272	2.6278	2.5241	2.4151	2.2997	2.1760
40	3.1167	2.9531	2.7811	2.5984	2.5020	2.4015	2.2958	2.1838	2.0635	1.9318
60	2.9042	2.7419	2.5705	2.3872	2.2898	2.1874	2.0789	1.9622	1.8341	1.6885
120	2.7052	2.5439	2.3727	2.1881	2.0890	1.9839	1.8709	1.7469	1.6055	1.4311
∞	2.5188	2.3583	2.1868	1.9998	1.8983	1.7891	1.6691	1.5325	1.3637	1.0000

TABLE A5 Critical values of S for the Wilcoxon matched-pairs signed-rank test

Two-sided	$n = 5$	$n = 6$	$n = 7$	$n = 8$	$n = 9$	$n = 10$
$P = .10$	1	2	4	6	8	11
$P = .05$		1	2	4	6	8
$P = .02$			0	2	3	5
$P = .01$				0	2	3

Two-sided	$n = 11$	$n = 12$	$n = 13$	$n = 14$	$n = 15$	$n = 16$
$P = .10$	14	17	21	26	30	36
$P = .05$	11	14	17	21	25	30
$P = .02$	7	10	13	16	20	24
$P = .01$	5	7	10	13	16	19

Two-sided	$n = 17$	$n = 18$	$n = 19$	$n = 20$	$n = 21$	$n = 22$
$P = .10$	41	47	54	60	68	75
$P = .05$	35	40	46	52	59	66
$P = .02$	28	33	38	43	49	56
$P = .01$	23	28	32	37	43	49

Two-sided	$n = 23$	$n = 24$	$n = 25$	$n = 26$	$n = 27$	$n = 28$
$P = .10$	83	92	101	110	120	130
$P = .05$	73	81	90	98	107	117
$P = .02$	62	69	77	85	93	102
$P = .01$	55	61	68	76	84	92

Two-sided	$n = 29$	$n = 30$	$n = 31$	$n = 32$	$n = 33$	$n = 34$
$P = .10$	141	152	163	175	188	201
$P = .05$	127	137	148	159	171	183
$P = .02$	111	120	130	141	151	162
$P = .01$	100	109	118	128	138	149

Two-sided	$n = 35$	$n = 36$	$n = 37$	$n = 38$	$n = 39$	
$P = .10$	214	228	242	256	271	
$P = .05$	195	208	222	235	250	
$P = .02$	174	186	198	211	224	
$P = .01$	160	171	183	195	208	

Two-sided	$n = 40$	$n = 41$	$n = 42$	$n = 43$	$n = 44$	$n = 45$
$P = .10$	287	303	319	336	353	371
$P = .05$	264	279	295	311	327	344
$P = .02$	238	252	267	281	297	313
$P = .01$	221	234	248	262	277	292

Two-sided	$n = 46$	$n = 47$	$n = 48$	$n = 49$	$n = 50$	
$P = .10$	389	408	427	446	466	
$P = .05$	361	379	397	415	434	
$P = .02$	329	345	362	380	398	
$P = .01$	307	323	339	356	373	

For $n > 50$, use the large sample method described in Sec. 7.9.

Adapted from *Some Rapid Approximate Statistical Procedures* by Wilcoxon and Wilcoxon, 1964, with permission by Lederle Laboratories, a division of the American Cyanamid Company, Pearl River, N.Y.

TABLE A6 Critical values of R for the two-sample rank test

n_1	n_2	Two-Sided P 0.10	Two-Sided P 0.05	Two-Sided P 0.01	n_1	n_2	Two-Sided P 0.10	Two-Sided P 0.05	Two-Sided P 0.01
2	4	—	—	—	4	16	24;60	21;63	15;69
	5	3;13	—	—		17	25;63	21;67	16;72
	6	3;15	—	—		18	26;66	22;70	16;76
	7	3;17	—	—		19	27;69	23;73	17;79
	8	4;18	3;19	—		20	28;72	24;76	18;82
	9	4;20	3;21	—	5	5	19;36	17;38	15;40
	10	4;22	3;23	—		6	20;40	18;42	16;44
	11	4;24	3;25	—		7	21;44	20;45	16;49
	12	5;25	4;26	—		8	23;47	21;49	17;53
	13	5;27	4;28	—		9	24;51	22;53	18;57
	14	6;28	4;30	—		10	26;54	23;57	19;61
	15	6;30	4;32	—		11	27;58	24;61	20;65
	16	6;32	4;34	—		12	28;62	26;64	21;69
	17	6;34	5;35	—		13	30;65	27;68	22;73
	18	7;35	5;37	—		14	31;69	28;72	22;78
	19	7;37	5;39	3;41		15	33;72	29;76	23;82
	20	7;39	5;41	3;43		16	34;76	30;80	24;86
3	3	6;15	—	—		17	35;80	32;83	25;90
	4	6;18	—	—		18	37;83	33;87	26;94
	5	7;20	6;21	—		19	38;87	34;91	27;98
	6	8;22	7;23	—		20	40;90	35;95	28;102
	7	8;25	7;26	—	6	6	28;50	26;52	23;55
	8	9;27	8;28	—		7	29;55	27;57	24;60
	9	10;29	8;31	6;33		8	31;59	29;61	25;65
	10	10;32	9;33	6;36		9	33;63	31;65	26;70
	11	11;34	9;36	6;39		10	35;67	32;70	27;75
	12	11;37	10;38	7;41		11	37;71	34;74	28;80
	13	12;39	10;41	7;44		12	38;76	35;79	30;84
	14	13;41	11;43	7;47		13	40;80	37;83	31;89
	15	13;44	11;46	8;49		14	42;84	38;88	32;94
	16	14;46	12;48	8;52		15	44;88	40;92	33;99
	17	15;48	12;51	8;55		16	46;92	42;96	34;104
	18	15;51	13;53	8;58		17	47;97	43;101	36;108
	19	16;53	13;56	9;60		18	49;101	45;105	37;113
	20	17;55	14;58	9;63		19	51;105	46;110	38;118
4	4	11;25	10;26	—		20	53;109	48;114	39;123
	5	12;28	11;29	—	7	7	39;66	36;69	32;73
	6	13;31	12;32	10;34		8	41;71	38;74	34;78
	7	14;34	13;35	10;38		9	43;76	40;79	35;84
	8	15;37	14;38	11;41		10	45;81	42;84	37;89
	9	16;40	14;42	11;45		11	47;86	44;89	38;95
	10	17;43	15;45	12;48		12	49;91	46;94	40;100
	11	18;46	16;48	12;52		13	52;95	48;99	41;106
	12	19;49	17;51	13;55		14	54;100	50;104	43;111
	13	20;52	18;54	13;59		15	56;105	52;109	44;117
	14	21;55	19;57	14;62		16	58;110	54;114	46;122
	15	22;58	20;60	15;65					

Reproduced with permission from *Elementary Medical Statistics*, 2nd ed., by Mainland, 1963, by W. B. Saunders Company, Philadelphia, Pennsylvania. The table was prepared by M. I. Sutcliffe.

n_1	n_2	Two-Sided P			n_1	n_2	Two-Sided P		
		0.10	0.05	0.01			0.10	0.05	0.01
7	17	61;114	56;119	47;128	12	12	120;180	115;185	105;195
	18	63;119	58;124	49;133		13	125;187	119;193	109;203
	19	65;124	60;129	50;139		14	129;195	123;201	112;212
	20	67;129	62;134	52;144		15	133;203	127;209	115;221
						16	138;210	131;217	119;229
8	8	51;85	49;87	43;93					
	9	54;90	51;93	45;99		17	142;218	135;225	122;238
	10	56;96	53;99	47;105		18	146;226	139;233	125;247
	11	59;101	55;105	49;111		19	150;234	143;241	129;255
	12	62;106	58;110	51;117		20	155;241	147;249	132;264
	13	64;112	60;116	53;123	13	13	142;209	136;215	125;226
	14	67;117	62;122	54;130		14	147;217	141;223	129;235
	15	69;123	65;127	56;136		15	152;225	145;232	133;244
	16	72;128	67;133	58;142		16	156;234	150;240	136;254
	17	75;133	70;138	60;148		17	161;242	154;249	140;263
	18	77;139	72;144	62;154		18	166;250	158;258	144;272
	19	80;144	74;150	64;160		19	171;258	163;266	147;282
	20	83;149	77;155	66;166		20	175;267	167;275	151;291
9	9	66;105	62;109	56;115	14	14	166;240	160;246	147;259
	10	69;111	65;115	58;122		15	171;249	164;256	151;269
	11	72;117	68;121	61;128		16	176;258	169;265	155;279
	12	75;123	71;127	63;135		17	182;266	172;276	159;289
	13	78;129	73;134	65;142		18	187;275	179;283	163;299
	14	81;135	76;140	67;149		19	192;284	183;293	168;308
	15	84;141	79;146	69;156		20	197;293	188;302	172;318
	16	87;147	82;152	72;162					
	17	90;153	84;159	74;169	15	15	192;273	184;281	171;294
	18	93;159	87;165	76;176		16	197;283	190;290	175;305
						17	203;292	195;300	180;315
	19	96;165	90;171	78;183		18	208;302	200;310	184;326
	20	99;171	93;177	81;189		19	214;311	205;320	189;336
10	10	82;128	78;132	71;139		20	220;320	210;330	193;347
	11	86;134	81;139	73;147					
	12	89;141	84;146	76;154	16	16	219;309	211;317	196;332
	13	92;148	88;152	79;161		17	225;319	217;327	201;343
	14	96;154	91;159	81;169		18	231;329	222;338	206;354
						19	237;339	228;348	210;366
	15	99;161	94;166	84;176		20	243;349	234;358	215;377
	16	103;167	97;173	86;184					
	17	106;174	100;180	89;191	17	17	249;346	240;355	223;372
	18	110;180	103;187	92;198		18	255;357	246;366	228;384
	19	113;187	107;193	94;206		19	262;367	252;377	234;395
						20	268;378	258;388	239;407
	20	117;193	110;200	97;213					
					18	18	280;386	270;396	252;414
11	11	100;153	96;157	87;166		19	287;397	277;407	258;426
	12	104;160	99;165	90;174		20	294;408	283;419	263;439
	13	108;167	103;172	93;182					
	14	112;174	106;180	96;190	19	19	313;428	303;438	283;458
	15	116;181	110;187	99;198		20	320;440	309;451	289;471
	16	120;188	113;195	102;206	20	20	348;472	337;483	315;505
	17	123;196	117;202	105;214					
	18	127;203	121;209	108;222					
	19	131;210	124;217	111;230					
	20	135;217	128;224	114;238					

For $n_1 > 20$ and $n_2 > 20$, use the large sample method described in Sec. 7.8.

TABLE A7 95% confidence interval (percent) for binomial distribution

Number Observed x	Size of Sample, n												Fraction Observed x/n	Size of Sample			
	10		15		20		30		50		100*			250†		1000†	
0	0	31	0	22	0	17	0	12	0	07	0	4	.00	0	1	0	0
1	0	45	0	32	0	25	0	17	0	11	0	5	.01	0	4	0	2
2	3	56	2	40	1	31	1	22	0	14	0	7	.02	1	5	1	3
3	7	65	4	48	3	38	2	27	1	17	1	8	.03	1	6	2	4
4	12	74	8	55	6	44	4	31	2	19	1	10	.04	2	7	3	5
5	19	81	12	62	9	49	6	35	3	22	2	11	.05	3	9	4	7
6	26	88	16	68	12	54	8	39	5	24	2	12	.06	3	10	5	8
7	35	93	21	73	15	59	10	43	6	27	3	14	.07	4	11	6	9
8	44	97	27	79	19	64	12	46	7	29	4	15	.08	5	12	6	10
9	55	100	32	84	23	68	15	50	9	31	4	16	.09	6	13	7	11
10	69	100	38	88	27	73	17	53	10	34	5	18	.10	7	14	8	12
11			45	92	32	77	20	56	12	36	5	19	.11	7	16	9	13
12			52	96	36	81	23	60	13	38	6	20	.12	8	17	10	14
13			60	98	41	85	25	63	15	41	7	21	.13	9	18	11	15
14			68	100	46	88	28	66	16	43	8	22	.14	10	19	12	16
15			78	100	51	91	31	69	18	44	9	24	.15	10	20	13	17
16					56	94	34	72	20	46	9	25	.16	11	21	14	18
17					62	97	37	75	21	48	10	26	.17	12	22	15	19
18					69	99	40	77	23	50	11	27	.18	13	23	16	21
19					75	100	44	80	25	53	12	28	.19	14	24	17	22
20					83	100	47	83	27	55	13	29	.20	15	26	18	23
21							50	85	28	57	14	30	.21	16	27	19	24
22							54	88	30	59	14	31	.22	17	28	19	25
23							57	90	32	61	15	32	.23	18	29	20	26
24							61	92	34	63	16	33	.24	19	30	21	27
25							65	94	36	64	17	35	.25	20	31	22	28
26							69	96	37	66	18	36	.26	20	32	23	29
27							73	98	39	68	19	37	.27	21	33	24	30
28							78	99	41	70	19	38	.28	22	34	25	31
29							83	100	43	72	20	39	.29	23	35	26	32
30							88	100	45	73	21	40	.30	24	36	27	33
31									47	75	22	41	.31	25	37	28	34
32									50	77	23	42	.32	26	38	29	35
33									52	79	24	43	.33	27	39	30	36
34									54	80	25	44	.34	28	40	31	37
35									56	82	26	45	.35	29	41	32	38
36									57	84	27	46	.36	30	42	33	39
37									59	85	28	47	.37	31	43	34	40
38									62	87	28	48	.38	32	44	35	41
39									64	88	29	49	.39	33	45	36	42
40									66	90	30	50	.40	34	46	37	43
41									69	91	31	51	.41	35	47	38	44
42									71	93	32	52	.42	36	48	39	45
43									73	94	33	53	.43	37	49	40	46
44									76	95	34	54	.44	38	50	41	47
45									78	97	35	55	.45	39	51	42	48
46									81	98	36	56	.46	40	52	43	49
47									83	99	37	57	.47	41	53	44	50
48									86	100	38	58	.48	42	54	45	51
49									89	100	39	59	.49	43	55	46	52
50									93	100	40	60	.50	44	56	47	53

*If x exceeds 50, read $100 - x$ = number observed and subtract each confidence limit from 100.

†If x/n exceeds 0.50, read $1.00 - x/n$ = fraction observed and subtract each confidence limit from 100.

Reproduced by permission from *Statistical Methods*, 6th ed., by George W. Snedecor and William G. Cochran, © 1967 by the Iowa State University Press, Ames, Iowa.

TABLE A8 99% confidence interval (percent) for binomial distribution

Number Observed x	Size of Sample, n 10	15	20	30	50	100*	Fraction Observed x/n	Size of Sample 250†	1000†
0	0 41	0 30	0 23	0 16	0 10	0 5	.00	0 2	0 1
1	0 54	0 40	0 32	0 22	0 14	0 7	.01	0 5	0 2
2	1 65	1 49	1 39	0 28	0 17	0 9	.02	1 6	1 3
3	4 74	2 56	2 45	1 32	1 20	0 10	.03	1 7	2 4
4	8 81	5 63	4 51	3 36	1 23	1 12	.04	2 9	3 6
5	13 87	8 69	6 56	4 40	2 26	1 13	.05	2 10	3 7
6	19 92	12 74	8 61	6 44	3 29	2 14	.06	3 11	4 8
7	26 96	16 79	11 66	8 48	4 31	2 16	.07	3 13	5 9
8	35 99	21 84	15 70	10 52	6 33	3 17	.08	4 14	6 10
9	46 100	26 88	18 74	12 55	7 36	3 18	.09	5 15	7 12
10	59 100	31 92	22 78	14 58	8 38	4 19	.10	6 16	8 13
11		37 95	26 82	16 62	10 40	4 20	.11	6 17	9 14
12		44 98	30 85	18 65	11 43	5 21	.12	7 18	9 15
13		51 99	34 89	21 68	12 45	6 23	.13	8 19	10 16
14		60 100	39 92	24 71	14 47	6 24	.14	9 20	11 17
15		70 100	44 94	26 74	15 49	7 26	.15	9 22	12 18
16			49 96	29 76	17 51	8 27	.16	10 23	13 19
17			55 98	32 79	18 53	9 29	.17	11 24	14 20
18			61 99	35 82	20 55	9 30	.18	12 25	15 21
19			68 100	38 84	21 57	10 31	.19	13 26	16 22
20			77 100	42 86	23 59	11 32	.20	14 27	17 23
21				45 88	24 61	12 33	.21	15 28	18 24
22				48 90	26 63	12 34	.22	16 30	19 26
23				52 92	28 65	13 35	.23	17 31	20 27
24				56 94	29 67	14 36	.24	18 32	21 28
25				60 96	31 69	15 38	.25	18 33	22 29
26				64 97	33 71	16 39	.26	19 34	22 30
27				68 99	35 72	16 40	.27	20 35	23 31
28				72 100	37 74	17 41	.28	21 36	24 32
29				78 100	39 76	18 42	.29	22 37	25 33
30				84 100	41 77	19 43	.30	23 38	26 34
31					43 79	20 44	.31	24 39	27 35
32					45 80	21 45	.32	25 40	28 36
33					47 82	21 46	.33	26 41	29 37
34					49 83	22 47	.34	26 42	30 38
35					51 85	23 48	.35	27 43	31 39
36					53 86	24 49	.36	28 44	32 40
37					55 88	25 50	.37	29 45	33 41
38					57 89	26 51	.38	30 46	34 42
39					60 90	27 52	.39	31 47	35 43
40					62 92	28 53	.40	32 48	36 44
41					64 93	29 54	.41	33 50	37 45
42					67 94	29 55	.42	34 51	38 46
43					69 96	30 56	.43	35 52	39 47
44					71 97	31 57	.44	36 53	40 48
45					74 98	32 58	.45	37 54	41 49
46					77 99	33 59	.46	38 55	42 50
47					80 99	34 60	.47	39 55	43 51
48					83 100	35 61	.48	40 56	44 52
49					86 100	36 62	.49	41 57	45 53
50					90 100	37 63	.50	42 58	46 54

*If x exceeds 50, read $100 - x =$ number observed and subtract each confidence limit from 100.

†If x/n exceeds 0.50, read $1.00 - x/n =$ fraction observed and subtract each confidence limit from 100.

Reproduced by permission from *Statistical Methods*, 6th ed., by George W. Snedecor and William G. Cochran, © 1967 by the Iowa State University Press, Ames, Iowa.

TABLE A9   Critical values for the two-sided test that $\rho = 0$

n	.1	P .05	.01	.001	n	.1	P .05	.01	.001
5	.805	.878	.959	.991	30	.306	.361	.463	.570
6	.729	.811	.917	.974	35	.283	.334	.430	.532
7	.669	.754	.875	.951	40	.264	.312	.403	.501
8	.621	.707	.834	.925	45	.248	.294	.380	.474
9	.582	.666	.798	.898	50	.235	.279	.361	.451
10	.549	.632	.765	.872	55	.224	.266	.345	.432
11	.521	.602	.735	.847	60	.214	.254	.330	.414
12	.497	.576	.708	.823	65	.206	.244	.317	.399
13	.476	.553	.684	.801	70	.198	.235	.306	.385
14	.457	.532	.661	.780	80	.185	.220	.286	.361
15	.441	.514	.641	.760	90	.175	.207	.270	.341
16	.426	.497	.623	.742	100	.165	.197	.257	.324
17	.412	.482	.606	.725	120	.151	.179	.234	.297
18	.400	.468	.590	.708	140	.140	.166	.217	.275
19	.389	.456	.575	.693	160	.131	.155	.203	.258
20	.378	.444	.561	.679	180	.123	.146	.192	.243
22	.360	.423	.537	.652	200	.117	.139	.182	.231
24	.344	.404	.515	.629	250	.104	.124	.163	.207
26	.330	.388	.496	.607	500	.074	.088	.115	.147
28	.317	.374	.479	.588	1000	.052	.062	.081	.104

For given sample size n, reject the hypothesis of zero correlation if $|r|$ is greater than the critical value. For one-sided test, the P value is halved and for $H_1: \rho < 0$ reject if r is less than the negative of the critical value and against $H_1: \rho > 0$ reject if r is greater than the critical value.

TABLE A10 95% confidence interval for ρ

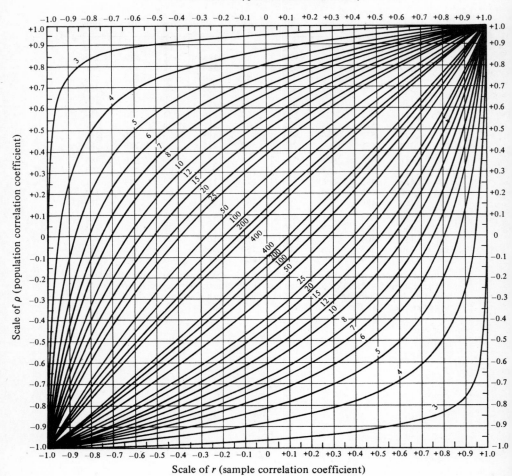

Scale of r (sample correlation coefficient)

The numbers on the curves indicate sample size. The chart can also be used to determine upper and lower 2.5% significance points for r, given ρ.

TABLE A11 99% confidence interval for ρ

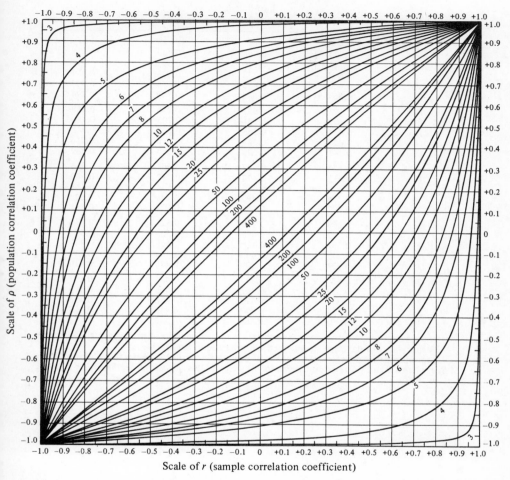

Scale of ρ (population correlation coefficient)

Scale of r (sample correlation coefficient)

The numbers on the curves indicate sample size. The chart can also be used to determine upper and lower 0.5% significance points for r, given ρ.

TABLE A12 Critical values of Spearman's rank correlation coefficient

		Two-Sided P		
n	.10	.05	.02	.01
5	0.900	—	—	—
6	0.829	0.886	0.943	—
7	0.714	0.786	0.893	—
8	0.643	0.738	0.833	0.881
9	0.600	0.683	0.783	0.833
10	0.564	0.648	0.745	0.794
11	0.523	0.623	0.736	0.818
12	0.497	0.591	0.703	0.780
13	0.475	0.566	0.673	0.745
14	0.457	0.545	0.646	0.716
15	0.441	0.525	0.623	0.689
16	0.425	0.507	0.601	0.666
17	0.412	0.490	0.582	0.645
18	0.399	0.476	0.564	0.625
19	0.388	0.462	0.549	0.608
20	0.377	0.450	0.534	0.591
21	0.368	0.438	0.521	0.576
22	0.359	0.428	0.508	0.562
23	0.351	0.418	0.496	0.549
24	0.343	0.409	0.485	0.537
25	0.336	0.400	0.475	0.526
26	0.329	0.392	0.465	0.515
27	0.323	0.385	0.456	0.505
28	0.317	0.377	0.448	0.496
29	0.311	0.370	0.440	0.487
30	0.305	0.364	0.432	0.478

For $n > 30$, use the large sample method described in Sec. 8.11.

Reproduced with permission of The Institute of Mathematical Statistics from "Critical Values of Spearman's Rank Correlation Coefficient," by E. G. Olds, *Annals of Mathematical Statistics*, Vol. 9, 1938.

TABLE A13 Sample sizes required in comparing two proportions

Required sample sizes for each of two samples in a test of $p_1 = p_2$ at level of significance α to have power $1 - \beta$ for the alternative $p_2 = p_1 + d$.

$$\alpha = \beta = 0.01$$

d \ p_1	.05	.10	.15	.20	.25	.30	.35	.40	.45	.50
.05	1,275	1,968	2,574	3,093	3,526	3,873	4,132	4,306	4,392	4,392
.10	424	586	727	846	944	1,019	1,073	1,106	1,117	1,106
.15	231	299	357	405	443	472	491	501	501	491
.20	152	188	217	242	261	274	283	285	283	274
.25	110	131	149	163	173	180	183	183	180	173
.30	85	98	109	117	123	127	128	127	123	117
.35	68	76	84	89	92	94	94	92	89	84
.40	55	61	66	70	72	72	72	70	66	61
.45	46	50	54	56	57	57	56	54	50	46
.50	39	42	44	45	46	45	44	42	39	35
.55	33	35	37	37	37	37	35	33	30	
.60	28	30	31	31	31	30	28	26		
.65	24	26	26	26	26	24	23			
.70	21	22	22	22	21	20				
.75	18	19	19	18	17					
.80	16	16	16	15						
.85	13	13	13							
.90	11	11								
.95	9									

$$\alpha = \beta = 0.05$$

d \ p_1	.05	.10	.15	.20	.25	.30	.35	.40	.45	.50
.05	676	1,023	1,327	1,587	1,803	1,977	2,107	2,193	2,236	2,236
.10	231	313	383	443	492	530	557	573	578	573
.15	128	162	191	215	235	249	259	264	264	259
.20	86	103	118	131	140	147	151	152	151	147
.25	63	73	82	89	94	98	100	100	98	94
.30	49	55	61	65	68	70	71	70	68	65
.35	39	44	47	50	52	53	53	52	50	47
.40	32	35	38	40	41	41	41	40	38	35
.45	27	29	31	32	33	33	32	31	29	27
.50	23	25	26	27	27	27	26	25	23	21
.55	20	21	22	22	22	22	21	20	18	
.60	17	18	19	19	19	18	17	16		
.65	15	16	16	16	16	15	14			
.70	13	14	14	14	13	13				
.75	12	12	12	12	11					
.80	10	10	10	10						
.85	9	9	9							
.90	8	8								
.95	7									

TABLE A13 (Cont.) Sample sizes required in comparing two proportions

$$\alpha = \beta = 0.10$$

d \ p_1	.05	.10	.15	.20	.25	.30	.35	.40	.45	.50
.05	440	652	836	994	1,126	1,231	1,310	1,363	1,389	1,389
.10	155	205	248	284	314	337	353	353	366	363
.15	87	108	126	141	153	161	167	170	170	167
.20	59	70	79	87	83	97	99	100	99	97
.25	44	50	56	60	63	65	66	66	65	63
.30	34	38	42	44	46	47	48	47	46	44
.35	28	31	33	34	36	36	36	36	34	33
.40	23	25	27	28	28	28	28	28	27	25
.45	20	21	22	23	23	23	23	22	21	20
.50	17	18	19	19	19	19	19	18	17	16
.55	15	15	16	16	16	15	16	15	15	14
.60	13	13	14	14	14	13	13	12		
.65	11	12	12	12	12	11	11			
.70	10	10	10	10	10	10				
.75	9	9	9	9	9					
.80	8	8	8	8						
.85	7	7	7							
.90	6	6								
.95	5									

$$\alpha = \beta = .20$$

d \ p_1	.05	.10	.15	.20	.25	.30	.35	.40	.45	.50
.05	230	323	403	472	529	575	609	632	643	543
.10	86	108	127	143	156	166	174	178	179	178
.15	50	60	68	74	80	83	86	87	87	86
.20	35	40	44	47	50	52	53	53	53	52
.25	26	29	32	34	35	36	37	37	36	35
.30	21	23	24	26	27	27	27	27	27	26
.35	17	19	20	20	21	21	21	21	20	20
.40	15	16	16	17	17	17	17	17	16	16
.45	13	14	14	14	14	14	14	14	13	13
.50	11	11	12	12	12	12	12	11	11	10
.55	10	10	10	10	10	10	10	10	9	
.60	9	9	9	9	9	9	9	8		
.65	8	8	8	8	8	8	7			
.70	7	7	7	7	7	7				
.75	6	6	6	6	6					
.80	6	6	6	6						
.85	5	5	5							
.90	5	5								
.95	4									

TABLE A14   Values of $Z(r) = \frac{1}{2} \log \frac{1+r}{1-r}$

r	$Z(r)$	r	$Z(r)$	r	$Z(r)$
.00	.000				
.01	.010	.36	.377	.71	.887
.02	.020	.37	.388	.72	.908
.03	.030	.38	.400	.73	.929
.04	.040	.39	.412	.74	.950
.05	.050	.40	.424	.75	.973
.06	.060	.41	.436	.76	.996
.07	.070	.42	.448	.77	1.020
.08	.080	.43	.460	.78	1.045
.09	.090	.44	.472	.79	1.071
.10	.100	.45	.485	.80	1.099
.11	.110	.46	.497	.81	1.127
.12	.121	.47	.510	.82	1.157
.13	.131	.48	.523	.83	1.188
.14	.141	.49	.539	.84	1.221
.15	.151	.50	.549	.85	1.256
.16	.161	.51	.563	.86	1.293
.17	.172	.52	.576	.87	1.333
.18	.182	.53	.590	.88	1.376
.19	.192	.54	.604	.89	1.422
.20	.203	.55	.618	.90	1.472
.21	.213	.56	.633	.91	1.528
.22	.224	.57	.648	.92	1.589
.23	.234	.58	.662	.93	1.658
.24	.245	.59	.678	.94	1.738
.25	.255	.60	.693	.95	1.832
.26	.266	.61	.709	.96	1.946
.27	.277	.62	.725	.97	2.092
.28	.288	.63	.741	.98	2.298
.29	.299	.64	.758	.99	2.647
.30	.310	.65	.775		
.31	.321	.66	.793		
.32	.332	.67	.811		
.33	.343	.68	.829		
.34	.354	.69	.848		
.35	.365	.70	.867		

Answers to Selected Problems

Numerical answers may depend on the extent of rounding off errors, and the reader should not expect to agree precisely with all of the given answers.

Chapter 1

1. (a) $\frac{1}{6}$, (b) 0.5, (c) $\frac{1}{6}$
2. (a) 0.4, (b) 0.9, (c) 0.1
3. (a) 0.4, (b) 0.5, (c) $\frac{1}{3}$, (d) No, since $P(A \mid B) \neq P(A)$
4. 1.0
5. (a) $\frac{1}{3}$, (b) 0.9, (c) 0.5
6. (a) 0.1, (b) 0.9, (c) 0.2
7. $\frac{1}{24}$
9. $\frac{95}{99}$
11. 0.9
12. (a) 0.2 (b) 21%
13. (a) 0.026, (b) 0.798, (c) 0.483
15. 0.97
16. 0.84
17. 0.83
18. (a) 0.58, (b) 0.071 *9.429.*
19. (a) 0.50, (b) 0.05

Chapter 2

3. (a) 0.75, (b) 0.704, (c) 0.296 *.704*
4. (a) 0.75, (b) 0.471, (c) 0.293
5. (a) 0.75 and 0.188, (b) 3.0 and 6.0, (c) 1.0 and 0.667
7. 3.50 and 1.20
8. (a) 0.33, (b) 1.82 and 1.70, (c) 4.88 and 5.20
1.5676 & 2.31

9. 0, 0, and 0.408
11. (a) 0.90 vs. 0.81,   (b) 0.90 vs. 0.99,   (c) \$315 vs. \$315
13. (a) \$9629,   (b) 0.159

Chapter 3

1. (a) 0.027,   (b) 0.343,   (c) 0.973
3. 0.103   0.0103
4. (a) 0.998,   (b) 0.800   .544
5. (a) 0.192, [1] (b) 0.950   .9679
7. 5
8. 0.083
9. (a) 0.050,   (b) 0.136,   (c) 0.728
11. (a) 4.9%,   (b) 0.12%
12. 80.6
13. (a) $\binom{200}{100}(0.6)^{100}(0.4)^{100}$,   (b) less than 0.001
15. (a) 11.5%,   (b) 2.4%
16. 0.996
18. 0.143   .822
19. (a) 0.019,   (b) 0.159
21. (a) 0.008,   (b) 1.313 million dollars

Chapter 4

3. Random error, time difference, bias
5. (b) Lognormal distribution
7. $\bar{X} = 18,075$ and $s = 2907$
9. 0.99
15. 7.23 vs. 7.20

Chapter 5

1. (a) No,   (b) A,   (c) can't say
3. 1.07 to 1.33
4. 8.71 to 11.04
5. (a) 121.1 to 128.9
6. 0.558 to 0.842
8. $\bar{X}/(\bar{X} + 1)$
10. $\bar{X}$
11. $\hat{\mu} = \bar{X}$ and $\hat{\sigma}^2 = \sum (X_t - \bar{X})^2/n$
13. (a) No; $Z = -3.6$,   (b) 0.0005
17. 2.07%
19. (a) 0.1026,   (b) 0.0975
20. 0.24%
21. 62

Chapter 6

1. (a) $Z = -2.21$; hence, reject the hypothesis at $P < 0.05$, (b) 42.5 to 49.5%
3. $\chi^2 = 0.31$; hence, accept the goodness of fit.
4. $\chi^2 = 10.24$; hence, influence of payment is significant at $P < 0.005$.
7. (a) $\chi^2 = 12.68$; thus, the relation is significant at $P < 0.01$, (b) $\chi^2 = 9.74$; hence, significantly different at $P < 0.01$. a) $.001 < P < .005$
9. $\chi^2 = 0.95$; hence, difference is not significant.
11. $\chi^2 = 9.66$; hence, four brands are significantly different at $P < 0.05$.
12. $\chi^2 = 5.77$; hence, hypertension is independent of smoking. $.01 < P < .025$
15. (a) Binomial with $n = 3$ and $p = 0.5$, (b) $\chi^2 = 0.22$; hence, accept the goodness of fit.
16. (a) Yes, since $P > 0.1$, (b) no, $n = 2642$
17. (a) $Z = -1.47$; $P < 0.2$, (b) 892
18. (a) 100, (b) 363, (c) $\chi^2 = 6.07$; hence, significant at $P < 0.05$.

Chapter 7

1. (a) $\chi^2 = 28.75$; hence, reject at $P < 0.05$, (b) 2.15 to 3.98
3. (a) 39.7 to 45.7
5. (a) $t = -4.43$; hence, difference is significant at $P < 0.001$, (b) $R = 38.5$; hence, significant at $P < 0.01$.
6. (a) $t = -2.94$; hence, the change is significant at $P < 0.05$, (b) $S = 2.5$; hence, significant at $P < 0.01$.
8. $t = -3.67$; hence, difference is significant at $P < 0.002$. $.002 < P < .01$
9. $t = 5.61$; hence, difference is significant at $P < 0.001$.
11. (a) $t = 3.0$; hence, do not approve using one-sided test, (b) Yes, not approving possibly acceptable drug.
13. $t = 3.32$; hence, reject the hypothesis at $P < 0.01$.
15. Compare the mean height, the mean weight, the variances of height, and the variances of weight.
17. Not significant ($P = 0.07$)
19. $Z = (74 - 95)/(74 + 95)^{1/2} = -1.62$; hence, the difference is not significant.
21. About 25

Chapter 8

1. (a) $\hat{Y} = 42.5 - 0.457X$, (b) 3.71, (c) accept, (d) no
3. (a) Positive correlation, (b) very high positive correlation, (c) no correlation, (d) negative correlation, (e) no correlation
5. (a) 0.030, (b) 0.002
6. Correlation significant at $P < 0.001$
7. Yes, since $\chi^2 = 29.08$

9. (a) $r = 0.349$; not significant,   (b) -0.20 to 0.71

6. (a) Partial correlation coefficient $r = 0.147$; significant at $P < 0.05$, (b) $r = 0.022$; hence, not significant.

17. Partial correlation coefficient between yield and temperature is 0.097; hence, although not significant, they are positively correlated if the effect of rainfall is removed.

19. $Z = 0.88$; hence, not significant.

20. $r_s = 0.958$; hence, significant at $P < 0.05$.

Chapter 9

1. (a)

Source of variation	S.S.	D.F.	M.S.	F	
Angle	93.13	2	46.56	68.51	$P < 0.001$
Within levels	5.44	8	0.68		

(b) All three means are different from each other at $P < 0.001$.

3.

Source of variation	S.S.	D.F.	M.S.	F	
Temperature	10502.04	1	10502.04	291.08	$P < 0.001$
Oven	6310.15	2	3155.07	87.45	$P < 0.001$
Interaction	72.17	2	36.08	0.60	N.S.
Within cells	360.50	6	60.08		

(Model II is assumed)

5.

Source of variation	S.S.	D.F.	M.S.	F	
Chemical	0.56	1	0.56	0.69	N.S.
Fertilizer	113.19	3	37.73	46.58	$P < 0.001$
Interaction	45.69	3	15.23	18.80	$P < 0.001$
Within cells	6.50	8	0.81		

(Model I is assumed)

7.

Source of variation	S.S.	D.F.	M.S.	F	
Week	1463.06	1	1463.06	9.92	$P < 0.05$
Diet	189.86	3	63.29	0.40	N.S.
Interaction	470.79	3	156.93	1.06	N.S.
Within cells	885.19	6	147.53		

(Mixed model with the diets being fixed)

9.

Source of variation	S.S.	D.F.	M.S.	F	
Group	1.68	4	0.420	10.50	$P < 0.001$
Within group	2.22	55	0.040		

The analysis shows that the overall difference is significant at $P < 0.001$. The LSD test indicates that the mean protein level for each of the first two groups is significantly different from each of the other three groups at $P < 0.001$.

Index

Testing the null hypothesis about the difference between two population means

$H_0: \mu_d = d_0 \qquad H_1: \mu_1 - \mu_2 = d_0$

— paired differences (dependent samples)

$$t = \frac{\bar{d} - d_0}{s_d/\sqrt{n}}$$

$$DF = n - 1 = \text{no. of pairs} - 1$$

— pop. variances known (independent samples)

$$Z = \frac{(\bar{X}_1 - \bar{X}_2) - d_0}{\sqrt{\dfrac{\sigma_1^2}{n_1} + \dfrac{\sigma_2^2}{n_2}}}$$

— pop. variances are unknown
but $\sigma_1 = \sigma_2$

$$t = \frac{(\bar{X}_1 - \bar{X}_2) - d_0}{S_P\sqrt{1/n_1 + 1/n_2}}$$

$$S_P = \sqrt{\frac{(n_1-1)S_1^2 + (n_2-1)S_2^2}{n_1 + n_2 - 2}} \quad \text{—— DF}$$

— pop. variances are unknown & unequal

$$t = \frac{(\bar{X}_1 - \bar{X}_2) - d_0}{\sqrt{\dfrac{S_1^2}{n_1} + \dfrac{S_2^2}{n_2}}}$$

$$d.f. = \frac{\left(\dfrac{S_1^2}{n_1} + \dfrac{S_2^2}{n_2}\right)^2}{\dfrac{\left(\dfrac{S_1^2}{n_1}\right)^2}{n_1+1} + \dfrac{\left(\dfrac{S_2^2}{n_2}\right)^2}{n_2+2}}$$